33⁹⁵

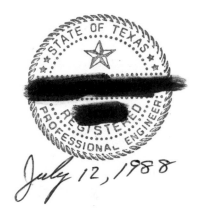

July 12, 1988

Simplified Design
of Reinforced Concrete

||||||||||||||||||||||||||||||||||||||||||

Simplified Design
of Reinforced Concrete

||||||||||||||||||||||||||||||||||||||||||||||||||||||||||||||||

The Late Harry Parker, M.S.
Formerly Professor of Architectural Construction
University of Pennsylvania

FIFTH EDITION

prepared by
JAMES AMBROSE, M.S.
Professor of Architecture
University of Southern California

A Wiley-Interscience Publication
JOHN WILEY & SONS
New York • Chichester • Brisbane • Toronto • Singapore

Copyright © 1984 by John Wiley & Sons, Inc.

All rights reserved. Published simultaneously in Canada.

Reproduction or translation of any part of this work
beyond that permitted by Section 107 or 108 of the
1976 United States Copyright Act without the permission
of the copyright owner is unlawful. Requests for
permission or further information should be addressed to
the Permissions Department, John Wiley & Sons, Inc.

Library of Congress Cataloging in Publication Data:

Parker, Harry, 1887–
 Simplified design of reinforced concrete.
 "A Wiley-Interscience publication."
 Bibliography: p.
 Includes index.
 1. Reinforced concrete construction. I. Ambrose,
James E. II. Title.
 TA683.2.P3 1984   693'.54   84-10462
 ISBN 0-471-80349-9

Printed in the United States of America

10 9 8 7 6 5 4 3 2 1

Preface to the Fifth Edition

||||||||||||||||||||||||||||||||||||||||||||||||

The publication of this edition presents the opportunity for yet another generation to have access to the late Professor Parker's enduringly popular work. With the seemingly ever-increasing sophistication and complexity of codes and practices, the need for a brief, simplified explanation of basic methods of design is, if anything, even greater than it was at the time of publication of the first edition. While retaining Professor Parker's basic intentions and style, this edition brings the material into general conformity with present technology and practices. The work continues to offer illustrations of the practical design of ordinary elements of reinforced concrete, while the text material remains accessible to a person with a limited background in mathematics and engineering fundamentals.

Although present codes and most practicing engineers favor exclusive use of strength design methods, the old working stress method is still allowed in limited form by inclusion in an appendix to the ACI Code. For those preparing for practice as professional engineers, a thorough study of strength design is strongly recommended. However, for most situations, the working stress method is much simpler and easier for the less well trained to follow. For this reason the working stress method has been used in this edition wherever it is still permitted by the present code. For most problems, however, the strength method is also explained; and for some situations—such as column design and bar development—where the present code does not permit use of the working stress method, only strength design methods are used.

Some new material has been added in this edition, including

illustrations of the design of pedestals, short cantilever retaining walls, freestanding walls, and continuous span slabs and beams. Some materials included in previous editions (e.g., treatments of design of prestressed concrete and flat slab systems) has been omitted, since an effective treatment by present standards now seems to be beyond the scope of this type of book.

A major revision in this edition is the inclusion of data and computations in the metric-based Système International (SI) units. While work in the building field continues to utilize the old English (now more properly called U.S.) units, it is likely that persons now beginning to become involved in this field will need to develop the facility to work in both unit systems. Work in this edition is developed primarily in the traditional units, but wherever feasible, the corresponding material has been included in SI units. This presents the opportunity for the reader to learn whichever of the two systems he or she is least familiar with, utilizing the work as a running translation.

I am grateful to the American Concrete Institute for permission to use material from *Building Code Requirements for Reinforced Concrete* (known as the ACI Code), which remains the most widely used design standard in the United States. I am also grateful to the staff at John Wiley & Sons for their work in production and their general support. Finally, I am once again indebted to my family for their patience, indulgence, and general assistance.

JAMES AMBROSE

Westlake Village, California
August 1984

Preface to the First Edition
IIIIIIIIIIIIIIIIIIIIIIIIIIIIIIIIIIIIIIIIIIIII

The preparation of this book has been prompted by the fact that many young men desirous of the ability to design elementary reinforced concrete structural members have been deprived of the usual preliminary training. The author has endeavored to simplify the subject matter for those having a minimum of preparation. Throughout the text will be found references to Section I of *Simplified Engineering for Architects and Builders*. Familiarity with this brief treatment of the principles of mechanics is sufficient. Any textbook on mechanics will give the desired information. This particular book has been referred to as a convenience in having a direct reference. With these basic principles, and a high school knowledge of algebra, no other preparation is needed.

In preparing material for this book, the author has had in mind its use as a textbook as well as a book to be used for home study. Simple, concise explanations of the design of the most common structural elements have been given rather than discussions of the more involved problems. In addition to the usual design formulas sufficient theory underlying the principles of design is presented, in developing basic formulas, to ensure the student a thorough knowledge of the fundamentals involved.

A major portion of the book contains illustrative examples giving the solution of the design of structural members. Accompanying the examples are problems to be solved by the student.

The usual tables necessary in the design of reinforced concrete are included. No supplementary books, tables, or charts are required. Where practicable, safe load tables have been added, but in each instance illustrative examples give the design steps show-

ing the underlying principles by means of which the table was prepared.

The titles of other volumes in this series of elementary books relating to structural design have included the words "for Architects and Builders." The purpose of this has been to convey the idea that the books are limited in scope and that they are not the comprehensive and thorough treatises demanded by engineers. It is found, however, that these books have a much wider use than was anticipated. Because of this, it has seemed advisable to omit the words "for Architects and Builders" from the title of the present volume even though it has been prepared with this particular group in mind.

Grateful appreciation is extended to the Portland Cement Association and the American Concrete Institute for their kindness and cooperation in granting permission to reproduce data and tables from their publications.

The author has made no attempt to offer short-cuts or originality in design. Instead, he has endeavored to present clearly and concisely the present day methods commonly used in the design of reinforced concrete members. A thorough knowledge of the principles herein set forth should encourage the student and serve as adequate preparation for advanced study.

HARRY PARKER

High Hollow, Southampton, Pa.
March 1943

Contents

IIIIIIIIIIIIIIIIIIIIIIIIIIIIIIIIIIIIIIIIIIIIIII

1   Introduction   1

1-1   Reinforced Concrete   1
1-2   Work in This Book   2
1-3   Reference Sources   2
1-4   Design Methods   3
1-5   Computations   4
1-6   Units of Measurement   4
1-7   Nomenclature   7
1-8   Symbols   8

**2   Materials and Properties of Reinforced
       Concrete   9**

2-1   Concrete   9
2-2   Cement   10
2-3   Mixing Water   11
2-4   Aggregates   11
2-5   Lightweight Aggregates   12
2-6   Admixtures   13
2-7   Air-Entrained Concrete   13
2-8   Steel Reinforcement   14
2-9   Strength of Concrete   15
2-10 Modulus of Elasticity   16
2-11 Creep of Concrete   17
2-12 Durability   17

2-13 Workability   18
2-14 Water–Cement Ratio   19
2-15 Watertightness   19
2-16 Proportioning Concrete Mixes   20

3   Production of Reinforced Concrete   **23**

3-1   General   23
3-2   Design and Construction Controls   24
3-3   Formwork   25
3-4   Placement   26
3-5   Laitance   27
3-6   Curing Concrete   27
3-7   Degrees of Exposure   28
3-8   Testing Concrete   29
3-9   Slump Test   29
3-10 Compression Test   31
3-11 Installation of Reinforcement   32

4   Investigation of Reinforced Concrete   **34**

4-1   Introduction   34
4-2   Investigation of Beams   34
4-3   Effects of Beam End Restraint   36
4-4   Effects of Concentrated Loads   39
4-5   Multiple-Beam Spans   40
4-6   Complex Loading and Span Conditions   40
4-7   Approximate Analysis of Continuous Frames   43
4-8   The Working Stress Method   43
4-9   The Strength Design Method   44

5   General Requirements of Reinforced
        Concrete Structures   **46**

5-1   Introduction   46
5-2   Cover of Reinforcement   46

5-3   Spacing of Reinforcement   47
5-4   Bending of Reinforcement   47
5-5   Minimum Dimensions for Concrete Members   48
5-6   Shrinkage and Temperature Reinforcement   50
5-7   Minimum Reinforcing   50

6   Flexural Behavior of Beams   52

6-1   Internal Resisting Moment   52
6-2   Distribution of Compressive Stress   54
6-3   Design Methods and Codes   55
6-4   Flexure Formulas: Working Stress Design   56
6-5   Use of Working Stress Formulas   60
6-6   Flexure Formulas: Strength Design   65
6-7   Use of Strength Design Formulas   68

7   Shear and Diagonal Tension   73

7-1   Shear Situations in Concrete Structures   73
7-2   Development of Shear Stress   74
7-3   Shear in Beams   76
7-4   Beam Shear: General Design Considerations   80
7-5   Design of Shear Reinforcement: Working
        Stress Method   83
7-6   Design of Shear Reinforcement: Strength
        Design   90

8   Development of Reinforcement   94

8-1   Bond Stress   94
8-2   Development of Reinforcement   96
8-3   Development of Tensile Reinforcement   96
8-4   Use of Hooked Ends   97

8-5   Development of Compressive Reinforcement   101
8-6   Bar Development in Simple Beams   101
8-7   Bar Development in Continuous Beams   105
8-8   Bar Development in Columns   107
8-9   Splices   108

9   Design of Beams   110

9-1   General   110
9-2   Beam Shape   111
9-3   Beam Width   112
9-4   Beam Depth   113
9-5   Deflection Control   114
9-6   Design Procedure for Rectangular Beams   118
9-7   Design Procedure for T-Beams   126
9-8   One-Way Solid Slabs   131
9-9   Shrinkage and Temperature Reinforcement for
      Slabs   132
9-10 Design of a One-Way Solid Slab   133
9-11 Beams with Compressive Reinforcement   137

10   Reinforced Concrete Floor Systems   147

10-1   Introduction   147
10-2   Dead Load   150
10-3   Floor Live Loads   150
10-4   One-Way Solid Slab and Beam Systems   152
10-5   Design of a Continuous One-Way Solid Slab   155
10-6   Design of a Continuous Beam   159
10-7   Design of a Continuous Girder   167
10-8   Concrete One-Way Joist Construction   171
10-9   Concrete Waffle Construction   174
10-10 Two-Way Spanning Solid Slab Construction   177
10-11 Use of Design Aids   181

11   Reinforced Concrete Columns   182

11-1   Introduction   182
11-2   Columns with Axial Load plus Bending
        Moment   182
11-3   Types of Reinforced Concrete Columns   184
11-4   General Requirements for Reinforced Concrete
        Columns   185
11-5   Design of Tied Columns   186
11-6   Bar Layouts for Rectangular Tied Columns   191
11-7   Design of Round Columns   194
11-8   Slenderness Effects in Compression Members   194

12   Footings   195

12-1   Building Foundations   195
12-2   Bearing Pressure for Design   195
12-3   Independent Column Footings   197
12-4   Design of a Column Footing   198
12-5   Load Tables for Column Footings   203
12-6   Pedestals   208
12-7   Wall Footings   214
12-8   Design of a Wall Footing   216
12-9   Load Tables for Wall Footings   221

13   Walls   222

13-1   Introduction   222
13-2   General Requirements for Reinforced Concrete
        Walls   223
13-3   Bearing Walls   224
13-4   Basement Walls   227

13-5 Retaining Walls   232
13-6 Shear Walls   241
13-7 Freestanding Walls   244

References   246

Index   247

Simplified Design
of Reinforced Concrete
||||||||||||||||||||||||||||||||||||||

1

Introduction

||||||||||||||||||||||||||||||||||||||||||||||||

1-1. Reinforced Concrete

Concrete, made from natural materials, was used by many ancient builders thousands of years ago. Modern concrete, made with industrially produced cement, was first produced in the early part of the nineteenth century, when the material known as *portland cement* was developed. Because of its lack of tensile strength, however, concrete was used principally for crude, massive structures—foundations, bridge abutments, and very thick walls.

In the late nineteenth century, several builders developed the technique of inserting iron or steel rods into relatively thin structures of concrete in order to enhance their ability to develop tension forces. This was the beginning of what we now know as *reinforced concrete*. Many of the types of construction developed by these early experimenters are still part of our technical vocabulary for building structures.

Today, the use of concrete is still expanding to ever more sophisticated applications—thin shells and folded plates, precast and prestressed systems, fiber-reinforced elements, and so on. However, the simple, basic elements and systems developed by the early experimenters—reinforced footings, slabs, beams, and

columns—continue in widespread use; essentially unchanged except for the continuing improvement of material quality and production techniques. The material in this book deals with these commonly used elements and systems of ordinary reinforced concrete.

1-2. Work in This Book

As the range of concrete applications has broadened, and our knowledge base increased as the result of experience, research, and study, the procedures for structural investigation and design of reinforced concrete have become increasingly complex. Adding to the complexity are the facts that the material is composite (interacting concrete and steel) and the structural systems tend to be highly indeterminate due to the typical continuous, monolithic nature of most construction. While attempting to utilize the latest codes and design practices, the work in this book is developed in a manner that keeps it within the range of accessibility for persons without extensive training in mathematics and engineering. In general, the mathematics used is within the scope of most first year high school algebra courses. In order to be able to begin the work at a reasonable level, however, it is assumed that the reader has had an introduction to the basic principles of engineering mechanics: statics and strength of materials. If this is not the case, the reader is advised to spend some time studying these basic principles before attempting to follow the work in this book.

For persons who need an introduction—or possibly a review—of the type described, there are several books that contain this material. For those with a limited background in mathematics it is probably advisable to look for a book that is written with this type of reader in mind. Two such books are *Simplified Mechanics and Strength of Materials,* 3rd ed., by Harry Parker, and *Simplified Engineering for Architects and Builders,* 6th ed., by Harry Parker (both published by John Wiley & Sons).

1-3. Reference Sources

There are many sources for information useful to the designer of reinforced concrete structures. These include texts, handbooks,

institute publications, research reports, periodicals, and product brochures from manufacturers. The sources used for the work in this book are listed in the References section at the back of the book.

The principal reference for construction requirements and design practices for concrete structures for buildings is *Building Code Requirements for Reinforced Concrete*—ACI Standard 318-77, published by the American Concrete Institute. This publication is commonly known, and hereinafter referred to as the ACI Code. As with all such standards, this document is revised periodically, and care should be taken to use the latest edition when doing any real design work.

All the reference materials required for the example computations and the exercise problems in this book are contained herein. However, the reader is advised to obtain a copy of the latest ACI Code, and possibly one or more of the available handbooks, for more complete reference for any actual design work.

1-4. Design Methods

The design of reinforced concrete structural members may be accomplished by two different methods. The first method is called working stress design; the second is called strength design. Most of the material in the present ACI Code (1977 ed.) is based on strength design. A limited treatment of working stress design is given in Appendix B of the code, under the title Alternate Design Method. The last edition of the ACI Code to present a full development of the working stress method was the 1963 edition (ACI 318-63). The development of the working stress method in this book is based primarily on the 1963 code, but also complies in general with the requirements for the alternate method presented in the 1977 code.

For the design of some structural members, either method will produce reasonably adequate results. For some situations, however, the working stress method is no longer considered proper, and strength design is strongly advised. These situations are discussed in the appropriate sections of the book.

1-5. Computations

In most professional design firms structural computations are most commonly done with computers, particularly when the work is complex or repetitive. Anyone aspiring to participation in professional design work is advised to acquire the background and experience necessary to the application of computer-aided techniques. The computational work in this book is simple and can be performed easily with a pocket calculator. The reader who has not already done so is advised to obtain one. The "pocket slide rule" type with eight-digit capacity is quite sufficient.

For the most part, structural computations can be rounded off. Accuracy beyond the third place is seldom significant, and this is the level used in this work. In some examples more accuracy is carried in early stages of the computation to ensure the desired degree in the final answer. All the work in this book, however, was performed on an eight-digit pocket calculator.

1-6. Units of Measurement

At the time of preparation of this edition the building industry in the United States is still in a state of confused transition from the use of English units (feet, pounds, etc.) and the new metric-based system referred to as the SI units (for Système International). Although a complete phase-over to SI units seems inevitable, at the time of this writing the construction materials and products suppliers in the United States are still resisting it. Consequently, the AISC Manual and most building codes and other widely used references are still in the old units. (The old system is now more appropriately called the U.S. system, because England no longer uses it!) Although it results in some degree of clumsiness in the work, we have chosen to give the data and computations in this book in both units as much as is practicable. The technique is generally to perform the work in U.S. units and immediately follow it with the equivalent work in SI units enclosed in brackets [thus] for separation and identity.

Table 1-1 lists the standard units of measurement in the U.S.

TABLE 1-1. Units of Measurement: U.S. System

Name of unit	Abbreviation	Use
Length		
Foot	ft	large dimensions, building plans, beam spans
Inch	in.	small dimensions, size of m cross sections
Area		
Square feet	ft^2	large areas
Square inches	in.^{2}	small areas, properties of cross sections
Volume		
Cubic feet	ft^3	large volumes, quantities of materials
Cubic inches	in.^{3}	small volumes
Force, mass		
Pound	lb	specific weight, force, load
Kip	k	1000 pounds
Pounds per foot	lb/ft	linear load (as on a beam)
Kips per foot	k/ft	linear load (as on a beam)
Pounds per square foot	lb/ft^2, psf	distributed load on a surface
Kips per square foot	k/ft^2, ksf	distributed load on a surface
Pounds per cubic foot	lb/ft^3, pcf	relative density, weight
Moment		
Foot-pounds	ft-lb	rotational or bending moment
Inch-pounds	in.-lb	rotational or bending moment
Kip-feet	k-ft	rotational or bending moment
Kip-inches	k-in.	rotational or bending moment
Stress		
Pounds per square foot	lb/ft^2, psf	soil pressure
Pounds per square inch	lb/in.^{2}, psi	stresses in structures
Kips per square foot	k/ft^2, ksf	soil pressure
Kips per square inch	k/in.^{2}, ksi	stresses in structures
Temperature		
Degree Fahrenheit	°F	temperature

TABLE 1-2. Units of Measurement: SI System

Name of unit	Abbreviation	Use
Length		
Meter	m	large dimensions, building plans, beam spans
Millimeter	mm	small dimensions, size of member cross sections
Area		
Square meters	m^2	large areas
Square millimeters	mm^2	small areas, properties of cross sections
Volume		
Cubic meters	m^3	large volumes
Cubic millimeters	mm^3	small volumes
Mass		
Kilogram	kg	mass of materials (equivalent to weight in U.S. system)
Kilograms per cubic meter	kg/m^3	density
Force (load on structures)		
Newton	N	force or load
Kilonewton	kN	1000 newtons
Stress		
Pascal	Pa	stress or pressure (1 pascal = 1 N/m^2)
Kilopascal	kPa	1000 pascals
Megapascal	MPa	1,000,000 pascals
Gigapascal	GPa	1,000,000,000 pascals
Temperature		
Degree Celsius	°C	temperature

system with the abbreviations used in this work and a description of the type of use in structural work. In similar form Table 1-2 gives the corresponding units in the SI system. For more ready access the conversion units used in shifting from one system to the other appear on the inside back cover of this book.

1-7. Nomenclature

Notation used in this book complies with that used in the 1977 ACI Code. The following list includes all of the notation used in this book and is compiled and adapted from a more extensive list given in Appendixes B and C of the code.

A_c = area of concrete; gross area minus area of reinforcing

A_g = gross area of section ($A_c + A_s$)

A_s = area of reinforcing

A_s' = area of compressive reinforcement in a doubly reinforced section

A_v = area of shear reinforcing

A_1 = loaded area in bearing

A_2 = gross area of bearing support member

E_c = modulus of elasticity of concrete

E_s = modulus of elasticity of steel

M = design moment

N = design axial load

V = design shear force

a = depth of equivalent rectangular stress block (strength design)

b = width of compression face of member

b_w = width of stem in a T-beam

c = distance from extreme compression fiber to the neutral axis (strength design)

d = effective depth, from extreme compression fiber to centroid of tensile reinforcing

e = eccentricity of a nonaxial load, from the centroid of the section to the point of application of the load

f_c = unit compressive stress in concrete

f'_c = specified compressive strength of concrete

f_s = stress in reinforcement

f_y = specified yield stress of steel

h = overall thickness of member; unbraced height of a wall

jd = length of internal moment arm

kd = distance from extreme compression fiber to the neutral axis (working stress)

n = modular ratio of elasticity: E_s/E_c

p = percent of reinforcing with working strength design, expressed as a ratio: A_s/A_g

s = spacing of stirrups

t = thickness of a solid slab

ρ = percent of reinforcing with ultimate strength design expressed as a ratio: A_s/A_g

ϕ = strength reduction factor (strength design)

1-8. Symbols

The following "shorthand" symbols are frequently used.

Symbol	Reading
$>$	is greater than
$<$	is less than
$\geqq$	equal to or greater than
$\leqq$	equal to or less than
6'	6 feet
6"	6 inches
Σ	the sum of
ΔL	change in L

2

Materials and Properties of Reinforced Concrete
||||||||||||||||||||||||||||||||||||||||||||||||

2-1. Concrete

Concrete is produced by mixing a paste of cement and water with various inert materials. The most commonly used inert materials are sand and gravel or crushed stone. As soon as the cement and water are combined, a chemical process called hydration begins and in a few hours the mixture begins to harden, with the paste functioning to bind the inert materials into a solid mass. Under ideal conditions, the hardening process will continue until the concrete assumes a very hard, rocklike character.

The properties of the finished concrete, including its strength, weight, color, and porosity, are subject to considerable variation. Variables include the type of cement; the ratio of water to cement; the type, size, and proportionate amount of the inert materials; various actions performed while mixing and depositing the wet mix; and the conditions that occur while the concrete is hardening (called the curing period).

Well-made concrete has significant resistance to compressive stress, but is relatively weak in resisting tension. In some applications, concrete may be required to resist primarily compressive

9

stress and thus require no assistance. For most applications, however, some assistance must be provided to allow the development of tension resistance. The most common assistance is provided in the form of steel bars, which are placed in the concrete mass prior to hardening. When the finished structural element is loaded, the steel bars help to strengthen—or reinforce—the concrete.

2-2. Cement

The cement used most extensively in building construction is *portland cement*. Among the five types of standard portland cement generally available in the United States, and for which the American Society for Testing and Materials has established specifications, are two that account for most of the cement used in buildings: ASTM Type I, a general-purpose cement for use in concrete designed to reach its required strength in about 28 days, and ASTM Type III, a high-early strength cement for use in concrete that attains its design strength in a period of a week or less. All portland cements set and harden by reacting with water, and this hydration process is accompanied by the generation of heat. In massive concrete structures such as dams, the resulting temperature rise of the materials becomes a critical factor in both design and construction but the problem is not generally significant in building construction. (ASTM Type IV, a low-heat cement, is designed for use when heat rise during hydration is a critical factor.)

The controlling specifications for portland cement as called for in the ACI Code are the following:

1. *Specification for Portland Cement* (ASTM C150).
2. *Specification for Air-Entraining Portland Cement* (ASTM C175).

It is, of course, essential that the cement actually used in construction correspond to that employed in designing the mix to produce the specified compressive strength of the concrete.

High-early-strength cement is often used for construction during cold weather; the considerable heat developed during its rapid

gain in strength tends to prevent the concrete from freezing. It is also employed when the construction schedule calls for forms to be removed as soon as possible or when a structure must be placed in service in the shortest possible time.

2-3. Mixing Water

The water used in making concrete should be clean and free from injurious amounts of oil, acid, alkali, organic matter, or other deleterious substances. In general, any drinking water free from pronounced odor or taste is satisfactory for use as mixing water. Excessive impurities, however, may affect setting time and concrete strength and cause corrosion of reinforcement or efflorescence on finished concrete surfaces.

Although seawater containing as much as 3.5% salt can be used for making plain concrete, it should not be employed for reinforced concrete because of the risk of corrosion of the steel reinforcement.

2-4. Aggregates

The materials held together by the paste formed of cement and water are the *aggregates*. The aggregates are inert materials: natural sand, crushed stone, pebbles, cinders, and slag. The material smaller than $\frac{3}{8}$ in. in diameter is called *fine aggregate*. The fine aggregate should consist of natural sand, or of inert materials with similar characteristics, having clean, hard, and durable grains free from organic matter or loam. The size and grading of fine aggregate are determined by standard wire-cloth sieves. It is desirable to have a mixture of fine and coarse grains, for graded aggregate will produce a more compact, hence stronger, concrete. A common specification for grading fine aggregate requires that not less than 95 to 100% shall pass the No. 4 sieve and not more than 30 nor less than 10% shall pass the No. 50 sieve. Requirement for fine as well as coarse aggregate are given in *Specifications for Concrete Aggregates* (ASTM C33) and *Concrete Materials and Methods of Concrete Construction* (CSA 23.1). These specifications cover requirements for gradation,

abrasion resistance, and soundness, and limit the amounts of deleterious substances that may be present.

All material larger than $\frac{3}{8}$ in. in diameter is called *coarse aggregate* and includes crushed stone, gravel, slag, or other inert materials. Like the fine aggregate, coarse aggregate should also range in size. In general, the sizes vary from $\frac{1}{4}$ to 3 in., the maximum for reinforced concrete being 1 or $1\frac{1}{2}$ in. Some building codes limit the size of the coarse aggregate for reinforced concrete to three-quarters of the minimum clear spacing between reinforcing bars and not larger than one-fifth of the narrowest dimension between the sides of the forms of the member in which the concrete is to be used. When concrete members are small, necessitating the close spacing of bars, the coarse aggregate is usually graded $\frac{3}{8}$ to $\frac{3}{4}$ in. Any crushed rock of durable and strong qualities or clean hard gravel may be used as coarse aggregate. Trap rock makes one of the best aggregates. Granite and hard limestone are likewise suitable, but certain sandstones are considered unfit for use. In proportioning fine to coarse aggregate, there are, of course, many possible combinations, but, depending on its maximum size, the coarse aggregate will usually constitute approximately 50 to 60% of the total aggregate. The fine and coarse aggregates together generally occupy 60 to 80% of the volume of concrete.

2-5. Lightweight Aggregates

The aggregates discussed in the preceding section are those used in the production of normal-weight concrete, that is, concrete weighing about 135 to 160 lb/ft³. *Structural lightweight concretes* ranging from about 85 to 115 lb/ft³ are made with expanded shale, clay, slate, and slag as aggregates. Such materials produce concrete of sufficient strength for many purposes and, in comparison with stone concrete, reduce the dead loads appreciably. Other lightweight materials such as vermiculite and perlite are used to make *insulating concretes* (nonstructural) weighing about 15 to 30 lb/ft³. Structural lightweight aggregates should meet the provisions of *Specifications for Lightweight Aggregates for Structural Concrete* (ASTM C330), and aggregates for insulating concrete should conform to the requirements of *Specifications for Light-*

weight Aggregates for Insulating Concretes (ASTM C332). The design of structural lightweight concrete elements is not treated in this book.

2-6. Admixtures

Substances added to concrete to improve its workability, accelerate its set, harden its surface, and increase its waterproof qualities are known as *admixtures*. The term embraces all materials other than the cement, water, and aggregates that are added just before or during mixing. Many of the proprietary compounds contain hydrated lime, calcium chloride, and kaolin. Calcium chloride is the most commonly used admixture for accelerating the set of concrete, but corrosion of steel reinforcement may be the consequence of its excessive use. Caution should be exercised in the use of admixtures, especially those of unknown composition.

The desired properties produced by the use of admixtures in concrete can often be obtained just as economically and conveniently by proper proportioning of the mix and selection of suitable materials, without employing admixtures (except air-entraining admixtures as discussed in the following section).

2-7. Air-Entrained Concrete

Air-entrained concrete is produced by using an air-entraining portland cement (ASTM C175) or by introducing an air-entraining admixture as the concrete is mixed. In addition to improving workability, entrained air permits lower water–cement ratios (Section 2-9) and significantly improves the durability of hardened concrete. Air-entraining agents produce billions of microscopic air cells per cubic foot; they are distributed uniformly throughout the mass. These minute voids prevent the accumulation of water, which, on freezing, would expand and result in spalling of the exposed surface under frost action.

The use of entrained air is common in concrete for building construction. Although air-entraining cements reduce the strength of concrete somewhat and require slightly richer mix-

tures to obtain the same strength produced by normal portland cement, some reduction in strength is acceptable in view of the other favorable characteristics of air-entrained concrete. When admixtures are used to produce air entrainment, they should conform to *Specification for Air-Entraining Admixtures for Concrete* (ASTM C260).

2-8. Steel Reinforcement

The most common type of steel reinforcement employed in concrete building construction consists of round bars, usually of the deformed type, with lugs or projections on their surfaces. The purpose of the surface deformations is to develop a greater bond between the concrete and the steel. The bars used for reinforcement are made from billet steel, rail steel, or axle steel, conforming to ASTM Specifications A615, A616, and A617, respectively. The most common *grades* of reinforcing steel are Grade 40 and Grade 60, with yield strengths (f_y) of 40,000 and 60,000 lb/in.^{2} (psi), respectively.

Properties of standard steel reinforcing bars are listed in a table printed on the inside back cover of this book. Bar designation numbers are based on the number of eighths of an inch included in the nominal diameter of the bars.

Another type of reinforcement is *welded wire fabric,* which consists of a series of parallel longitudinal wires welded at regular intervals to transverse wires. It is available in sheets or rolls and is widely used as reinforcement in floor slabs and walls. The configuration of welded wire fabric achieves a more uniform distribution of steel by the use of smaller members more closely spaced than that provided by larger bars spaced more widely. Also, improved bond between the concrete and steel is obtained by the mechanical anchorage that results from the continuously welded transverse wires.

Welded wire fabric is made from cold-drawn wire, either smooth or deformed; the former has a yield strength of 65,000 or 56,000 psi, depending on the wire size, whereas the yield strength of the deformed wire is 70,000 psi.

A new method of designating wire sizes has superseded the

steel wire gage system once employed. The cross-sectional area of the wire is the basic element in the new system. Deformed wire sizes are specified by the letter D, followed by a number indicating hundredths of a square inch. Thus D20 designates a deformed wire with a cross-sectional area of 0.20 in.^{2}. Smooth wire sizes are similarly indicated by substituting W for D. The term *style* is used to identify the spacings and sizes of the wires in welded wire fabric. A typical style designation is 4 × 8—W16 × W10, which denotes a welded smooth wire fabric in which

spacing of longitudinal wires = 4 in.
spacing of transverse wires = 8 in.
size of longitudinal wires   = W16 (0.16 in.^{2})
size of transverse wires   = W10 (0.10 in.^{2})

A welded deformed-wire fabric style would be indicated in the same manner by substituting D for W.

2-9. Strength of Concrete

The designer of a reinforced concrete building bases his computations on the use of concrete with a specified compressive strength (say 3000 psi) at the end of a 28-day curing period. The symbol for this specified strength is f_c'. Concretes of different strengths are produced by varying the proportions of cement, fine aggregate (sand), coarse aggregate, and water in the mix. The general theory in establishing the proportions of fine and coarse aggregates is that the voids in the coarse aggregate should be filled with the cement paste and fine aggregate.

Very little concrete is proportioned and mixed at the building site today. Central or ready-mixed concrete is used whenever it is available. Concrete mixed under controlled conditions at a central plant affords many advantages. It is delivered to the building site in a revolving mixer. The proportions of cement, aggregate, and water are maintained accurately and any desired strength may be ordered. The product thus provided is uniform in quality.

Concrete in its plastic condition cannot, of course, be tested for strength, and the customary procedure is to take samples as

the concrete is mixed. After curing, the samples are subjected to tests for compression. In addition to compressive stresses, concrete must resist shear (diagonal tension) and the bonding stresses that are present where the reinforcing steel comes in contact with the concrete. We can, of course, test concrete for these individual stresses, but the compression test (Section 3-10) serves as a measure of the other strengths.

2-10.   Modulus of Elasticity

We know from mechanics that within the elastic limit of a material stress is directly proportional to deformation (strain) and that *modulus of elasticity* is defined by the expression

$$E = \frac{\text{unit stress}}{\text{unit strain}}$$

It is a measure of the stiffness of a material or its resistance to deformation and is expressed in pounds per square inch (psi).

The modulus of elasticity E_c of hardened concrete depends on w, the weight of the concrete, and on f'_c, is compressive strength. Its value may be estimated from the expression $E_c = w^{1.5}33\sqrt{f'_c}$ for values of w between 90 and 155 lb/ft^3. For normal weight concrete (145 lb/ft^3) E_c may be considered equal to $57{,}000\sqrt{f'_c}$. Thus for a concrete with $f'_c = 4000$ psi

$$E_c = 57{,}000\sqrt{f'_c} = 57{,}000\sqrt{4000} = 3{,}640{,}000 \text{ psi}$$

Table 2-1 gives values of E_c for five different strength concretes based on this formula when $w = 145$ lb/ft^3.

In computations for the design of reinforced concrete structural members it is necessary that we know the ratio of the modulus of elasticity of steel to that of the concrete we are using. This is known as the *modular ratio* and is denoted by the symbol n; thus $n = E_s/E_c$. For steel reinforcement $E_s = 29{,}000{,}000$ psi.

Consider a concrete for which $f'_c = 3000$ psi and $w = 145$ lb/ft^3. From Table 2-1, we find that the modulus of elasticity is $3{,}150{,}000$ psi. Then

$$n = \frac{E_s}{E_c} = \frac{29{,}000{,}000}{3{,}150{,}000} = 9.2$$

TABLE 2-1.   Modulus of Elasticity of Normal Weight Concrete

Ultimate compressive strength at 28-day period f_c' (psi)	Modulus of elasticity of concrete E_c (psi)	$n = \dfrac{E_s}{E_c}$
2000	2,549,000	11.3
2500	2,880,000	10.1
3000	3,150,000	9.2
4000	3,640,000	8.0
5000	4,070,000	7.1

The values of n for concretes of other strengths are listed in Table 2-1.

2-11. Creep of Concrete

In loaded concrete members there is a tendency for deformations to increase with lapse of time even under constant load; this deformation is called *creep*. The effect of creep is equivalent to a decrease in the modulus of elasticity. The values of E_c given in Table 2-1 may therefore be used only for computation of deflections that occur immediately on application of the service load. Long-time deflections of slabs, beams, and other floor systems may be two or three times larger than the initial deflection. Control of initial deflections as well as those due to creep is considered in subsequent chapters.

2-12. Durability

The use of reinforced concrete for structural members of buildings has increased with amazing rapidity. Although older structures invariably proved to have adequate strength for the imposed loads, there are many instances in which insufficient attention was given to the *durability* of the concrete. As used in building construction, concrete may have various *degrees of exposure;* for instance, columns and girders on the exterior of a structure are subjected to atmospheric conditions to which interior members

are not exposed. Again, walls and piers subjected to alternate wetting and drying, or freezing and thawing, must necessarily be made of concrete designed to withstand such conditions. Thus it is seen that the designer of a reinforced concrete structure must, in determining the proper mix, bear in mind the degree of exposure as well as the strength of the concrete.

2-13. Workability

In addition to the above qualifications, concrete in its plastic condition must have a consistency that will permit it to be placed readily in the forms. This quality is known as *workability*. Different classes of work require different degrees of plasticity. The shape, width, and depth of forms and spaces between reinforcement are all factors that determine the required degree of workability. It might appear that varying the amount of water in the mix would permit any desired consistency. In former years this procedure of producing workable concrete frequently resulted in a mixture containing an excess amount of water, which, on hardening, produced a porous concrete of lower strength than desired. As a result of innumerable tests and experience, it has been found that the quantity of water in relation to the quantity of cement is a vital factor in determining the strength of concrete, and, this desired ratio having been established, the degree of plasticity is best obtained by regulating the proportions of the cement and water paste with the aggregate.

To produce a first quality concrete, the use of a mixing machine is essential. Thorough mixing not only tends to produce a concrete of uniform quality, but longer periods of mixing also increase the strength of the concrete and a greater degree of workability is effected.

Many types of portable mixers are obtainable today. Their capacities range from 3 ft^3 to 4 yd^3. The strength and quality of the concrete depend principally on the length of time the concrete remains in the mixer rather than on the speed of rotation. Concrete should never be mixed less than one minute, and a longer period is desirable when conditions permit. When concrete of

superior quality is desired for extreme exposure conditions or for watertightness, longer periods of mixing are advantageous.

Central-mixed or ready-mixed concrete is used whenever it is available. Certificates indicating the composition of each truck-load ensure compliance with specifications relating to strength.

2-14. Water–Cement Ratio

The most important factor affecting the strength of concrete is the *water–cement ratio*. This is expressed as the number of pounds of water per pound of cement used in the mix (or the number of gallons of water for each 94-lb bag of cement). The water–cement ratio determines the density of the cement paste, which in turn determines the strength, durability, and watertightness of the hardened concrete. The relationship is an inverse one; that is, lower values of the ratio produce higher strengths. We know that the freshly mixed concrete must be workable. It should be neither too dry nor too wet. If it is too dry, it is difficult to place in the forms and resists packing; the result is honeycombing. If the concrete is too wet, segregation of the ingredients may occur. To produce a workable concrete more water must be used than the amount required for hydration of the cement; consequently building codes specify the maximum permissible water–cement ratios for concretes of specified design strengths.

2-15. Watertightness

The quality of watertightness in concrete is of extreme importance. Certain structures such as tanks or basement walls, roof slabs, and floors below grade must, of course, be watertight to prevent passage of water from one side to the other. There is another important reason, however; disintegration may be physical or chemical, and deterioration of exposed concrete or steel reinforcement near the surface is due largely to the penetration of moisture.

Several factors enter into the production of watertight concrete. Obviously the aggregates must be nonporous, durable, and

well-graded materials. The concrete must be dense; that is, the water–cement ratio must be as low as possible, and it must be borne in mind that the mix must be workable and that the particles of aggregate must be completely bound together in the cement paste. It is customary to retain the ingredients in the mixing machine for a somewhat longer period if a complete incorporation of cement paste and aggregates is to be obtained. More precautions than usual should be taken in placing the concrete. This demands careful spading or vibration so that the reinforcement is completely encased and a uniform and dense exposed surface results.

The importance of proper curing of concrete, if it is to be watertight, cannot be overestimated. This is particularly true in the early stages of hardening. The exposed surfaces must be kept continuously damp so that a hard, dense surface will result to prevent checking and dusting.

2-16. Proportioning Concrete Mixes

A detailed discussion of proportioning the ingredients of concrete to achieve specified design compressive strengths is not given in this book. The brief presentation which follows is intended only to serve as background information. Readers interested in studying this aspect of concrete manufacture should consult Chapter 4 of the ACI Code and the following references:

1. *Recommended Practice for Selecting Proportions for Normal and Heavy Weight Concrete* (ACI 211.1-74), American Concrete Institute.
2. *Design and Control of Concrete Mixtures,* Portland Cement Association.

The first step in determining the proportions of the various ingredients for a concrete of desired design strength is to establish the water–cement ratio. The next step is to decide on the most economical combination of fine and coarse aggregates that will result in a concrete with a plasticity (consistency) that is workable.

As noted earlier, the general theory in establishing the proportions of the fine and coarse aggregates is that the voids in the

coarse aggregate should be filled with the cement paste and fine aggregate. The voids in coarse aggregate depend on the kind of material and its size. In general, the voids average slightly less than one-half the volume, and it is customary to use about one-half as much sand as the volume of crushed stone. We express the proportions in this sequence: cement, sand, and coarse aggregate; for instance, the mix may be $1:2:4$, $1:2\frac{1}{2}:5$, $1:3:6$. Often the fine and coarse aggregates are given as one figure, and a mix of $1:2:4$ may be expressed as $1:6$. The reason is that the sand should not always be one-half the volume of the crushed stone, for it may prove to be more economical to use a $1:2\frac{1}{2}:3\frac{1}{2}$ mix. This, however, is another way of expressing a $1:6$ proportion. Sand containing an average amount of moisture will bulk about 20%, and this fact should always be kept in mind in determining the proportions to be used. The sum of fine and coarse aggregates in proportion to the cement paste depends on the consistency required. In general, stiffer mixes are more economical with respect to the cost of materials, but if the mix is unusually dry the cost of placing it in the forms will be increased and care must be taken to avoid honeycombing. When the proportion of fine aggregate is increased, a smoother working concrete results, but this generally requires more cement paste and may not be economical.

Probably the commonest mix for average job conditions is 1 of cement to 5 of combined aggregates, with a water–cement ratio of $6\frac{1}{2}$ if non-air-entrained concrete is used. The $1:5$ mix may be $1:1\frac{3}{4}:3\frac{1}{4}$ or $1:2:3$. Specimens thus made will produce a concrete with an ultimate compressive strength of about 3000 psi.

If a concrete of greater strength is desired, or if the degree of exposure is more severe, the water–cement ratio should be reduced to produce a denser concrete.

When the concrete structure is of sufficient magnitude to warrant the expense involved, another method of determining proportions may be used. Certain data are given the contractor in the specifications, such as class of work, required strength, maximum water–cement ratio, maximum sizes of aggregate, and slump range. The contractor must, of course, use a concrete that is plastic and workable. Within the specified limitations, a series

of tests is made of various proportions, and a water–cement strength curve is established. This method of trial batches, based on the water–cement ratio, permits the contractor to produce most economically a concrete with the required qualities. Having determined the water–cement ratio in accordance with the desired strength and resistance to exposure, he or she then selects the most suitable combination of aggregates that will produce the required degree of workability.

A procedure for carrying out the process described is given in the references cited at the beginning of this section. The first one listed (ACI 211.1-74) also contains a table of concrete mixes for small jobs on which time and personnel are not available to determine proportions in accordance with the more detailed procedure.

3

Production of Reinforced Concrete
||||||||||||||||||||||||||||||||||||||||||||||||

3-1. General

Because concrete is a mixture in which a paste made of portland cement and water binds together fine and coarse particles of inert materials, known as aggregates, it is readily seen that by varying the proportions of the ingredients innumerable combinations are possible. These combinations result in concrete of different qualities. When the cement has hydrated, the plastic mass changes to a material resembling stone. This period of hardening is called *curing*, in which three things are required: time, favorable temperatures, and the continued presence of water.

To fulfill requirements it is essential that the hardened concrete have, above all else, *strength* and *durability*. In order that the concrete in its plastic form may be readily placed in the forms, another essential quality is *workability*. When watertightness is required, concrete must be *dense* and *uniform* in quality. Hence it is seen that in determining the various proportions of the mixture the designer must have in mind the purpose for which the concrete is to be used and the exposure to which it will be subjected. The following factors regulate the quality of the concrete: suitable

materials, correct proportions, proper methods of mixing and placing, and adequate protection during curing.

3-2. Design and Construction Controls

Structural designers ordinarily document their work in the form of a set of written computations. These computations will include a listing of design criteria: codes and standards used, concrete strength and steel type used, design loadings assumed, and so on. The computations are concluded by a listing of the design decision information: required shape and dimensions of concrete elements and the positions, number, and size of reinforcing bars. In most cases some sketches are used in the computations, to indicate the arrangement of reinforcing in member cross sections and the locations of bar cutoffs, extensions, bend points, and so on.

Structural computations are not ordinarily used to transmit information to the builder. For this purpose, it is ordinarily the practice to produce a set of contract documents: working drawings and specifications. Translation of the computations into the construction documents is normally done in the design office, and the designer should understand this process and be able to check the final form of the construction documents in order to assure that his or her design has been properly translated.

While the construction documents (if thoroughly executed) completely delineate the finished structure, the builders must usually produce a second set of documents that explain more directly to the work force how to make the concrete forms, provide falsework (the supports required during pouring and curing of the concrete), and fabricate and install the steel bars. Although the correctness of this translation is the responsibility of the builder, the designer should also verify the accuracy to avoid mistakes that will delay the construction.

At various stages of the construction, the adequacy of the work should be verified by the designer. The proper shape, details, and dimensions of forms, and the proper installation of reinforcing should be checked prior to pouring of the concrete. Installation in the forms of inserts for attachment, piping, electrical

conduit, blocking for duct, wiring, and piping chases, etc., should be inspected to assure that they do not critically reduce the structural capacity of affected members.

The quality of the finished concrete will be affected by many factors, as discussed later in this chapter. Control of these factors by the designer is done primarily by well-written specifications, but a little nagging during construction doesn't hurt.

3-3. Formwork

An inherent property of concrete is that it may be made in any shape. The wet mixture is placed in *forms* constructed of wood, metal, or other suitable material in which it hardens or sets. The forms must be put together with quality workmanship, holding to close dimensional tolerances. Formwork should be strong enough to support the weight of the concrete and rigid enough to maintain position and shape. In addition, formwork should be tight enough to prevent the seepage of water and designed to permit ready removal.

Timber used for forms is usually surfaced on the side that comes in contact with the concrete, and frequently is oiled or otherwise sealed. This fills the pores of the wood, reduces absorption of water from the concrete mixture, produces smoother concrete surfaces, and permits the form boards to be more easily removed.

Steel forms have the decided advantage of being more substantial if they are to be reused. Steel gives smoother surfaces to the concrete, although it is almost impossible to avoid showing the joints. For ribbed floors, metal pans and domes are used extensively, and columns, circular in cross section, are invariably made with metal forms.

Because the formwork for a concrete structure constitutes a considerable item in the cost of the completed structure, particular care should be exercised in its design. It is desirable to maintain a repetition of identical units so that the forms may be removed and reused at other locations with a minimum amount of labor.

There are no exact rules concerning the length of time the

forms should remain in place. Obviously they should not be removed until the concrete is strong enough to support its own weight in addition to any loads that may be placed on it. Also, too early removal of forms introduces the possibility of excessive deflections. Sometimes the side forms of beams are removed before the bottom forms. When this is done, posts or shoring are placed under the bottoms of the members to give additional support. This is called *reposting* or *reshoring*. The minimum period during which forms must remain in place before stripping is usually governed by the local building code.

3-4. Placement

The consistency of concrete should be such that a mass of uniform quality will result when it has been deposited in the forms. Remember that concrete in its plastic condition is in reality a paste in which the aggregates are mixed. Care should be exercised to prevent the particles of sand and stone from being separated from this paste, for such separation produces an inferior concrete. Factors that must be considered in preventing segregation of aggregates are transporting the concrete from the mixing machine to the forms, dropping the concrete from too great a height, and tamping or spading. Dropping the concrete more than 3 ft into the forms tends to permit the larger aggregate to work its way to a lower level, thus preventing a uniform quality.

Great care must be taken to see that the plastic concrete flows properly into all corners and angles of the forms and that the reinforcement is completely surrounded with concrete. When concrete is placed in the forms by means of chutes, it is important that long flows be avoided because there is a possibility that the large aggregate will separate from the other materials in the mix. If the use of long chutes is unavoidable, the concrete should be deposited into a hopper from which it is taken before being placed in the forms. The purpose of this procedure is to remix the materials and thereby correct any segregation.

Another cause of segregation is an excess amount of tamping, vibrating, or puddling in the forms. To avoid honeycombing a common procedure is to spade the concrete where it comes in

contact with the forms. In accomplishing this, it is advisable not to spade too vigorously, since this could cause the materials to become separated.

3-5. Laitance

When an excess amount of water is used, the concrete, on curing, will have a milky layer composed of cement and fine aggregate on the upper surface of the mass. This is called *laitance*. When evaporation of the excess water has been completed, this layer becomes weak, porous, and readily disintegrated and permits water to pass through under slight pressure. The removal of this thin layer before placing more concrete is of little benefit, for several inches of inferior concrete remain below. It is quite possible that the concrete at the bottom of the placing may be hard and dense. If the presence of laitance is detected, several inches below the uppermost surface should be removed if a durable and strong concrete is desired. Therefore, to avoid laitance, use only a concrete of the correct consistency and water–cement ratio.

3-6. Curing Concrete

Regardless of the care taken in proportioning, mixing, and placing, first-quality concrete can be obtained only when due consideration and provision are made for curing. The hardening of concrete is due to the chemical reaction between the water and cement. This hardening continues indefinitely as long as moisture is present and the temperatures are favorable. The initial set does not begin until two or three hours after the concrete has been mixed. During this interval moisture evaporates, particularly on the exposed surfaces, and unless provision is made to prevent the loss of moisture, the concrete will craze. A typical specification requires that the concrete be so protected that there is no loss of moisture from the surface for a period of seven days when normal portland cement is used and three days when the cement is of high early strength.

To prevent the loss of moisture during curing several methods may be employed. When hard enough to walk on, slabs may be

covered with burlap which is kept wet or with a suitable building paper with the edges pasted down. Another method is to cover the slabs with a 1-in. layer of wet sand or sawdust. Frequently a 6-in. layer of wet straw or hay is placed on the slabs. Another method sometimes resorted to is the continuous sprinkling of the exposed surfaces with water. The early removal of forms permits undue evaporation; hence the forms should be allowed to remain for as long a period as is practicable. In addition to strength and durability, controlled curing is one of the best precautions in making a watertight concrete.

The period of protection against evaporation of moisture varies with the type of structure and climatic conditions. Thin sections or concrete placed during hot weather require an increased period of protection.

Low temperatures during the period of curing produce concrete of lower strength than concrete cured at 72°F. Freezing of concrete before it has cured should never be permitted to occur, for the resulting concrete will be of poor quality and indeterminate strength.

Although special precautions are required, concrete work may be continued during severe weather conditions. To keep the concrete above freezing the materials may be heated before mixing or the concrete may be protected with suitable covers or kept in heated enclosures. If the weather is only moderately cold, heating the water used for mixing may be a sufficient precaution. In more severe weather, it may be necessary to heat both water and aggregates. The materials should never have a temperature exceeding 90°F when deposited.

One common method of protecting concrete is to cover it with a thick layer of straw and tarpaulins. Canvas enclosures heated by steam give excellent protection, since desirable temperatures may be maintained and the concrete is protected against drying out. If other heating devices are used, care should be exercised to see that moisture is not evaporated from the concrete.

3-7. Degrees of Exposure

Most structural members in reinforced concrete buildings are not exposed to the weather but exterior columns and spandrel beams

frequently are. Alternate cycles of freezing and thawing, wetting and drying, and prolonged periods of surface wetting diminish the durability of concrete. Consequently, the degree of exposure must be taken into account when designing the mix.

In climates in which freezing occurs, entrained air should be used in all exposed concrete. Sulfate-resisting cements (ASTM Types II and V) should be used where concrete is exposed to seawater or comes in contact with sulfate-bearing soils. When conditions of exposure are severe, a low water–cement ratio should be used even though strength requirements may be met with a higher value.

3-8. Testing Concrete

If the operation is of sufficient magnitude, concrete made of various proportions and with aggregates from the sources proposed for use on the job should be tested before construction of the building is started. In the usual procedure several combinations are tested by using at least four different water–cement ratios. The results of the tests are then plotted and the most economical mix that will produce a concrete of the desired strength, density, and workability is chosen. It is customary to continue testing the concrete during construction, particularly if there are changing weather conditions or if a change is made in the sources from which the aggregates are obtained.

The two most common tests of concrete are the slump test for determining the degree of workability of the fresh concrete and the compression test on cylinders of cured concrete to establish its strength. The effectiveness of these tests in quality control of concrete production depends on obtaining truly representative samples of fresh concrete and following standard procedures during testing. The American Society for Testing and Materials issues ASTM Standards covering sampling and testing, which are prescribed procedures under the ACI Code.

3-9. Slump Test

Although the terms *consistency* and *workability* are not strictly synonymous, they are closely related. Consistency may be

loosely defined as the wetness of the concrete mixture; it is an index of the ease with which concrete will flow during placement. A concrete is said to be workable if it is readily placed in the forms for which it is intended; for instance, a concrete of given consistency may be workable in large open forms but not in small forms containing numerous reinforcing bars. With this understanding, the slump test may be considered a measure of the workability of fresh concrete.

The equipment for making a slump test consists of a sheet metal truncated cone 12 in. high with a base diameter of 8 in. and a top diameter of 4 in. Both top and bottom are open. Handles are attached to the outside of the mold. When a test is made, freshly mixed concrete is placed in the mold in a stipulated number of layers and each is rodded separately a specified number of times with a steel rod. When the mold is filled, the top is leveled off and the mold lifted at once. The slumping action of the concrete is measured by taking the difference in height between the top of the mold and the top of the slumped mass of concrete (Fig. 3-1).

If the concrete settles 3 in., we say that the particular sample has a 3-in. slump. Thus the degree of consistency of the concrete is ascertained. Table 3-1 recommends slump ranges for various types of construction when vibration is used to consolidate the concrete in the forms.

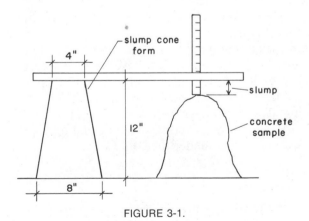

FIGURE 3-1.

TABLE 3-1. Recommended Slumps for Various Types of Construction[a]

	Slump (in.)	
Type of construction	Maximum[b]	Minimum
Reinforced foundation walls and footings	3	1
Plain footings, caissons, and substructure walls	3	1
Beams and reinforced walls	4	1
Building columns	4	1
Pavements and slabs	3	1
Mass concrete	2	1

[a] Data abstracted from *Recommended Practice for Selecting Proportions for Normal and Heavy Weight Concrete* (ACI 211.1-74) with permission of the American Concrete Institute.
[b] May be increased 1 in. for methods of consolidation other than vibration.

3-10. Compression Test

Tests of compressive strength are made at periods of 7 and 28 days on specimens prepared and cured in accordance with pre-scribed ASTM testing procedures. The specimen to be tested is cylindrical in shape and has a height twice its diameter. The standard cylinder is 6 in. in diameter and 12 in. high when the maximum size of the coarse aggregate does not exceed 2 in. For larger aggregates, the cylinder should have a diameter at least three times the maximum size of the aggregate and its height should be twice the diameter.

The mold used for the cylinders is made from metal or other nonabsorbent material such as paraffined cardboard. It is placed on a smooth plane surface (glass or metal plate) and filled with freshly made concrete in a specified number of layers. Each layer is consolidated by rodding or vibrating, either method being acceptable for concretes with a slump of 3 to 1 in. If the slump is greater than 3 in., the concrete must be rodded; if it is less than 1 in., it must be vibrated. As soon as casting of the cylinder is complete, the top of the specimen is covered to prevent the concrete from drying.

Because the strength of a specimen is greatly affected by tem-

perature changes, exposure to drying, and disturbances due to movement, it is customary to keep it at the site of operation for 24 hours. It is then taken to the laboratory and cured under controlled conditions in accordance with standard ASTM procedures. At the end of the curing period, each specimen is placed in the testing machine and a gradually increasing compressive load is applied until the specimen fails. The load causing failure is recorded, and this load, divided by the cross-sectional area of the cylinder, gives the ultimate compressive unit stress of the specimen. The same test is made on other specimens taken at the same time and cured under similar conditions, which, of course, results in a range of values for the compressive strength.

3-11. Installation of Reinforcement

To facilitate the shop fabrication and field installation of reinforcing bars, the bar supplier usually prepares a set of drawings—commonly called the *shop drawings*. These drawings consist of the supplier's interpretation of the engineering contract drawings, with the information necessary for the workers who fabricate the bars in the shop and those who install the bars in the field prior to pouring of the concrete. The exact cut lengths of bars, the location of all bends, the number of each type of bar, and so forth, will be indicated on these drawings. While the correctness of these drawings is the responsibility of the supplier, it is usually a good idea for the designer to verify the drawings in order to reduce mistakes in the construction.

Reinforcing bars must be held firmly in place during the pouring of the concrete. Horizontal bars must be held up above the forms; vertical bars must be braced from swaying against the forms. The positioning and holding of bars is done through the use of various accessories and a lot of light-gage tie wire. When concrete surfaces are to be exposed to view after being poured, it behooves the designer to be aware of the various problems of holding bars and bracing forms, since many of the accessories used ordinarily will be partly in view on the surface of the finished concrete.

Installation of reinforcing may be relatively simple and easy to

achieve, as in the case of a simple footing or a single beam. In other cases, where the reinforcing is extensive or complex, the problems of installation may require consideration during the design of the members. When beams intersect each other, or when beams intersect columns, the extended bars from the separate members must pass each other at the joint. Consideration of the "traffic" of the intersecting bars at such joints may affect the positioning of bars in the individual members.

4

Investigation of Reinforced Concrete

IIIIIIIIIIIIIIIIIIIIIIIIIIIIIIIIIIIIIIIIIIIIII

4-1. Introduction

Design of reinforced concrete structures often requires quite extensive and complex investigation of their behavior. The complexity of investigations is compounded by the use of the double material (concrete and steel), by the need to investigate many conditions for individual members, and the extensive continuity that exists in most poured-in-place concrete structures. The material in this chapter deals with the problems of determining the internal forces (axial loads, shears, and moments), of considerations for deflections, and of the investigation and design procedures for individual members.

4-2. Investigation of Beams

The simple, single-span beam is a rare situation in reinforced concrete structures. As shown in Fig. 4-1, the simple beam may exist when a single span is supported on bearing-type supports that offer little restraint (Fig. 4-1a), or when beams are connected to columns with connections that offer little moment resistance

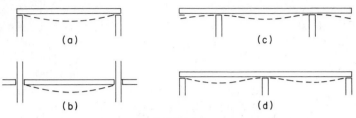

FIGURE 4-1.

(Fig. 4-1b). Although these situations are common in structures of steel and wood, they seldom occur in concrete structures, except when precast elements are used.

For single-story structures, supported on bearing-type supports, continuity resulting in complex bending can occur when the spanning members are extended over the supports. This may occur in the form of cantilevered ends (Fig. 4-1c), or of multiple spans (Fig. 4-1d). These conditions are common in wood and steel structures, and can also occur in reinforced concrete structures. Members of steel and wood are usually constant in cross section throughout their length; thus it is necessary only to find the single maximum value for shear and the single maximum value for moment. For the concrete member, however, the variations of shear and moment along the beam length must be considered, and several different cross sections must be investigated.

Figure 4-2 shows conditions that are common in concrete

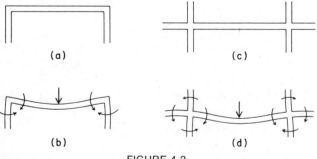

FIGURE 4-2.

structures when beams and columns are cast monolithically. For the single-story structure (Fig. 4-2a), the rigid joint between the beam and its supporting columns will result in behavior shown in Fig. 4-2b; with the columns offering some degree of restraint to the rotation of the beam ends. Thus some moment will be added to the tops of the columns and the beam will behave as for the center portion of the span in Fig. 4-1c, with both positive and negative moments.

For the multiple-story, multiple-span concrete frame, the typical behavior will be as shown in Figs. 4-2c and 4-2d. The columns above and below, plus the beams in adjacent spans, will contribute to the development of restraint for the ends of an individual beam span. This condition occurs in steel structures only when welded or heavily bolted moment-resisting connections are used. In concrete structures, it is the normal condition.

The structures shown in Figs. 4-1d, 4-2a, and 4-2c are statically indeterminate. This means that their investigation cannot be performed using only the conditions of static equilibrium. Although a complete consideration of statically indeterminant behaviors is well beyond the scope of this book, some treatment must be given for a realistic development of the topic of design of reinforced concrete structures. The discussions that follow will serve to illustrate the various factors in the behavior of continuous frames and will provide material for approximate analysis of common situations.

4-3. Effects of Beam End Restraint

Figures 4-3a to 4-3d show the effects of various end support conditions on a single-span beam with a uniformly distributed load. Similarly, Figures 4-3e to 4-3h show the conditions for a beam with a single concentrated load. Values are indicated for the maximum shears, moments, and deflection for each case. (Values for end reaction forces are not indicated, since they are the same as the end shears.)

We note the following for the four cases of end support conditions.

1. Figure 4-3 shows the cantilever beam, supported at only

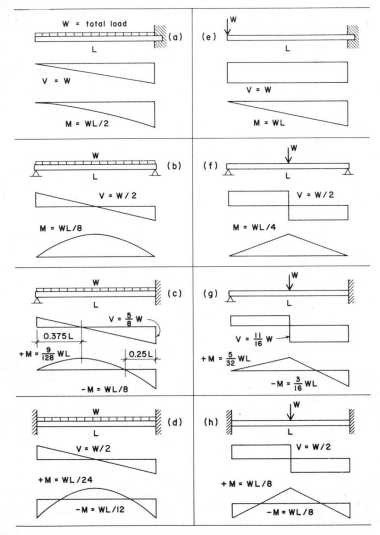

FIGURE 4-3. Beam response values for uniform and single concentrated loads.

one end with a *fixed end* condition. Both shear and moment are critical at the fixed end, and maximum deflection occurs at the unsupported end.

2. Figure 4-3*b* shows the classic "simple" beam, with supports offering only vertical force resistance. We will refer to this type of support as a *free end*. Shear is critical at the supports and both moment and deflection are maximum at the center of the span.

3. Figure 4-3*c* shows a beam with one free end and one fixed end. This support condition produces an unsymmetrical situation for the vertical reactions and the shear. The critical shear occurs at the fixed end, but both ends must be investigated separately for the concrete beam. Both positive and negative moments occur, with the maximum moment being the negative one at the fixed end. Maximum deflection will occur at some point slightly closer to the free end.

4. Figure 4-3*d* shows the beam with both ends fixed. This symmetrical support condition results in a symmetrical situation for the reactions, shear, and moments with the maximum deflection occurring at midspan. It may be noted that the shear diagram is the same as for the simple beam in Fig. 4-3*b*.

Continuity and end restraint have both positive and negative effects with regard to various considerations. The most positive gain is in the form of reduction of deflections, which is generally more significant for steel and wood structures, since deflections are less often critical for concrete members. For the beam with one fixed end (Fig. 4-3*c*), it may be noted that the value for maximum shear is increased and the maximum moment is the same as for the simple span (no gain in those regards). For full end fixity (Fig. 4-3*d*), the shear is unchanged, while both moment and deflection are quite substantially reduced in magnitude.

For the rigid frames shown in Fig. 4-2, the restraints will reduce moment and deflection for the beam, but the cost is at the expense of the columns, which must take some moment in addition to axial force. Rigid frames are often utilized to resist lateral loads due to wind and earthquakes, presenting complex combinations of lateral and gravity loading that must be investigated.

4-4. Effects of Concentrated Loads

Framing systems for roofs and floors often consist of series of evenly spaced beams that are supported by other beams placed at right angles to them. The supporting beams are thus subjected to a series of spaced, concentrated loads—the end reactions of the supported beams. The effects of a single such load at the center of a beam span are shown in Fig. 4-3e to 4-3h. Two additional situations of evenly spaced concentrated loading are shown in Fig. 4-4. When more than three such loads occur, it is usually ade-

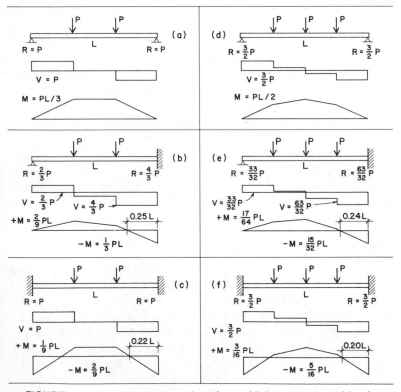

FIGURE 4-4.   Beam response values for multiple concentrated loads.

quate to consider the sum of the concentrated loads as a uniformly distributed load and to use the values given for Fig. 4-3*b* to 4-3*d*.

4-5. Multiple-Beam Spans

Figure 4-5 shows various loading conditions for a beam that is continuous through two equal spans. When continuous spans occur, it is usually necessary to give some consideration to the possibilities of partial beam loading, as shown in Figs. 4-5*b* and 4-5*d*. It may be noted for Fig. 4-5*b* that although there is less total load on the beam, the values for maximum positive moment, deflection, and shear at the free end are all higher than for the fully loaded beam in Fig. 4-5*a*. This condition of partial loading must be considered for *live loads* (people, furniture, snow, etc.). For design, the partial loading effects due to the live load must be combined with those produced by dead load (permanent weight of the construction) for the full action of the beam.

Figure 4-6 shows a beam that is continuous through three equal spans, with various situations of uniform load on the beam spans. Figure 4-6*a* gives the loading condition for dead load (*always* present in *all* spans). Figures 4-6*b* to 4-6*d* show the several possibilities for partial loading, each of which produces some specific critical values for the reactions, shears, moments, and deflections.

4-6. Complex Loading and Span Conditions

Although values have been given for many common situations in Figs. 4-3 to 4-6, there are numerous other possibilities in terms of unsymmetrical loadings, unequal spans, cantilevered free ends, and so on. Where these occur, an analysis of the indeterminate structure must be performed. For some additional conditions, the reader is referred to various handbooks that contain tabulations similar to those presented here. Two such references are the CRSI Manual (Ref. 3) and the AISC Manual (*Manual of Steel Construction,* 8th ed., published by the American Institute of Steel Construction).

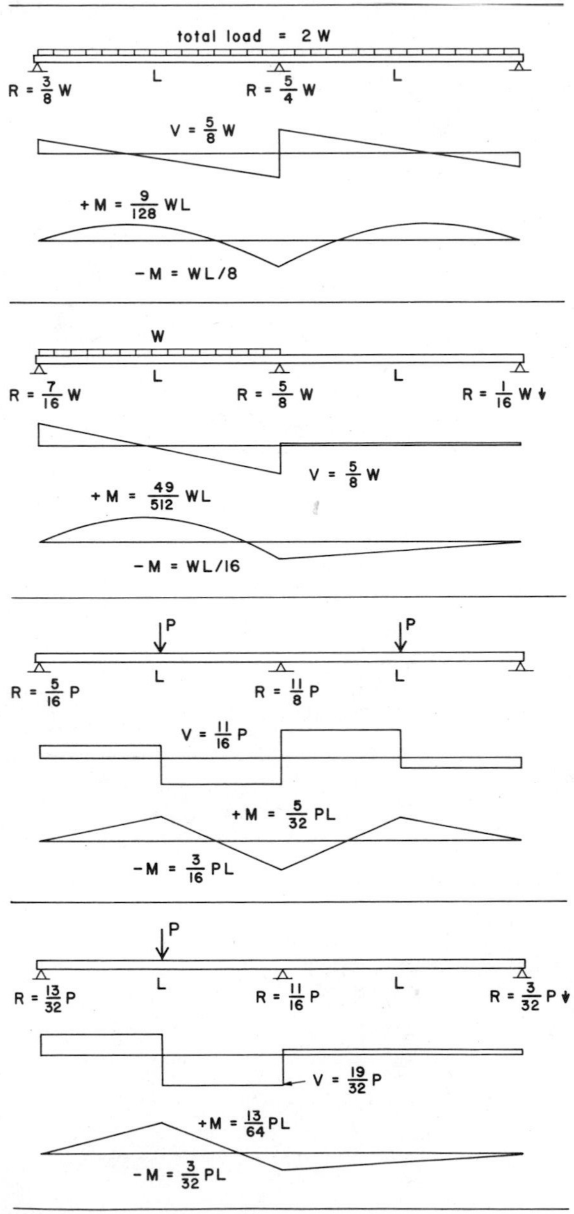
total load = 2 W
L
L
R = 3/8 W
R = 5/4 W
V = 5/8 W
+M = 9/128 WL
−M = WL/8
W
L
L
R = 7/16 W
R = 5/8 W
R = 1/16 W
V = 5/8 W
+M = 49/512 WL
−M = WL/16
P
P
L
L
R = 5/16 P
R = 11/8 P
V = 11/16 P
+M = 5/32 PL
−M = 3/16 PL
P
L
L
R = 13/32 P
R = 11/16 P
R = 3/32 P
V = 19/32 P
+M = 13/64 PL
−M = 3/32 PL

FIGURE 4-5. Response values for two-span beams.

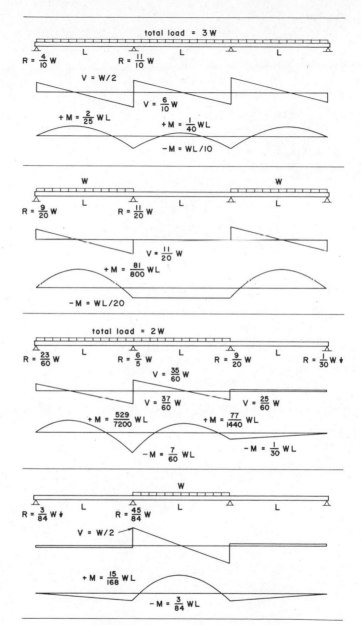

FIGURE 4-6. Response values for three-span beams.

4-7. Approximate Analysis of Continuous Frames

Because of the laborious nature of indeterminate structural analysis, and the many approximating assumptions that are ordinarily made in concrete design, the ACI Code permits the use of approximate coefficients for the determination of shears and moments for the design of slabs and beams under certain circumstances. The use of these approximate coefficients is illustrated in the sample computations in Chapter 10.

4-8. The Working Stress Method

As applied to the investigation of behaviors and the design of members of reinforced concrete, the working stress method consists of the determination of stresses in members that are induced by the actual loading under working conditions—called service load conditions. The stresses thus determined are then compared to the limits established for the situation under investigation. These limiting stresses—called allowable stresses—are established by code requirements as are the methods by which the actual stresses are determined. If the actual stresses do not exceed the allowable stresses, the member is considered to be adequate.

For concrete, allowable streeses are essentially based on the established design strength of the material. This strength is the so-called ultimate compressive strength, designated as f'_c, which is determined from the testing of standard samples, as discussed in Chapter 3.

Allowable stresses for steel are based on the yield strength of the steel, designated as f_y. Formulas used for determination of actual stresses in both the concrete and the steel are essentially derived from consideration of elastic behavior of members. In many cases, however, adjustments are made on the purely elastic formulas to account for the nonlinear stress-strain behavior of the concrete.

The working stress method is no longer favored by the codes and has largely been replaced in professional design practice by strength design methods. It is not our purpose to advocate the use

of one method over the other; both methods are presented for most of the work in this book. In general the working stress method is simpler to explain and its methods are easier and less complex to use. In some situations it is still used by designers and is allowed by most building codes. The latest edition of the ACI Code (1977 ed.) still provides for its use in a limited number of situations.

As applied to the work in this book, the working stress method is based primarily on the current requirements of the ACI Code. These are described in Appendix B of the code, under the title Alternate Method. The last edition of the code to develop the method in a full manner was the 1963 edition, and some of the material used here is taken from that publication.

4-9. The Strength Design Method

Application of the working stress method consists of designing members to *work* in an adequate manner (without exceeding established stress limits) under actual service load conditions. The basic procedure in strength design is to design members to *fail;* thus the ultimate strength of the member at failure (called its design strength) is the only type of resistance considered. Safety in strength design is not provided by limiting stresses, as in the working stress method, but by using a factored design load (called the *required strength*) that is greater than the service load. The code establishes the value of the required strength, called U, as not less than

$$U = 1.4D + 1.7L \tag{1}$$

in which D = the effect of dead load
L = the effect of live load

Other adjustment factors are provided when design conditions involve consideration of the effects of wind, earth pressure, differential settlement, creep, shrinkage, or temperature change.

The design strength of structural members (i.e., their *usable* ultimate strength) is determined by the application of assump-

tions and requirements given in the code and is further modified by the use of a *strength reduction factor* ϕ as follows:

ϕ = 0.90 for flexure, axial tension, and combinations of flexure and tension

    = 0.75 for columns with spirals

    = 0.70 for columns with ties

    = 0.85 for shear and torsion

    = 0.70 for compressive bearing

    = 0.65 for flexure in plain (not reinforced) concrete

Thus while Formula (1) may imply a relatively low safety factor, an additional margin of safety is provided by the stress reduction factors.

5

General Requirements of Reinforced Concrete Structures

||||||||||||||||||||||||||||||||||||||||||||||||||

5-1. Introduction

This chapter discusses some of the general requirements that apply to all concrete structures. Only those that apply to the structure as a whole are presented here. Requirements for particular types of members are given in the chapters that deal with those members.

5-2. Cover of Reinforcement

Steel bars are usually placed as close as possible to the outside surface of concrete members in order to be most effective in resisting flexure or to help in the reduction of surface cracking. The distance between the edge of the bars and the outside surface of the concrete is called the *cover*. General requirements for cover are as follows:

    1.   3 in. for the sides of members cast directly against soil.

2. For concrete exposed to soil or the weather and cast in forms: 2 in. for No. 6 bars and larger; 1.5 in. for No. 5 bars and smaller.
3. For concrete not exposed to weather or in contact with soil: 0.75 in. for No. 11 bars or smaller in slabs, walls, and joists; 1.5 in. in beams and columns.

5-3. Spacing of Reinforcement

Where multiple bars are used in members (which is the common situation), there are both upper and lower limits for the spacing of the bars. Lower limits are intended to permit adequate development of the concrete-to-steel stress transfers and to facilitate the flow of the wet concrete during pouring. For columns, the minimum clear distance between bars is specified as 1.5 times the bar diameter or a minimum of 1.5 in. For other situations, the minimum is one bar diameter or a minimum of 1 in.

For walls and slabs, maximum center-to-center bar spacing is specified as three times the wall or slab thickness or a maximum of 18 in. This applies to reinforcement required for computed stresses. For reinforcement that is required for control of cracking due to shrinkage or temperature change, the maximum spacing is five times the wall or slab thickness or a maximum of 18 in.

For adequate placement of the concrete, the largest size of the coarse aggregate should be not greater than three-quarters of the clear distance between bars.

5-4. Bending of Reinforcement

In various situations, it is sometimes necessary to bend reinforcing bars. Bending is done preferably in the fabricating shop instead of at the job site, and the bend diameter (see Fig. 5-1) should be adequate to avoid cracking the bar.

Bending of bars is sometimes done in order to provide anchorage for the bars. The code defines such a bend as a "standard hook," and the requirements for the details of this type of bend are given in Fig. 5-2.

As the yield stress of the steel is raised, bending becomes

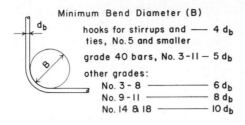

Minimum Bend Diameter (B)

hooks for stirrups and ——— $4\,d_b$
ties, No.5 and smaller

grade 40 bars, No. 3-11 — $5\,d_b$

other grades:

No. 3 - 8 ——————— $6\,d_b$
No. 9 - 11 —————— $8\,d_b$
No. 14 & 18 ————— $10\,d_b$

FIGURE 5-1.   Minimun bend requirements.

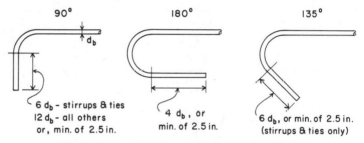

90°   180°   135°

d_b

$6\,d_b$ - stirrups & ties
$12\,d_b$ - all others
or, min. of 2.5 in.

$4\,d_b$, or
min. of 2.5 in.

$6\,d_b$, or min. of 2.5 in.
(stirrups & ties only)

FIGURE 5-2.   Requirements for standard hooks.

increasingly difficult. Bending of bars should be avoided when the yield stress exceeds 60 ksi [414 MPa]; and where it is necessary, should be done with bend diameters slightly greater than those given in Fig. 5-1.

5-5.   Minimum Dimensions for Concrete Members

For practical reasons, as well as the satisfying of various requirements for cover and bar spacing, there are minimum usable dimensions for various reinforced concrete members. When flexural reinforcement is required in slabs, walls, or beams, its effectiveness will be determined in part by the distance between the tension-carrying steel and the far edge of the compression-carrying concrete. Thus extremely shallow beams and thin slabs or walls will have reduced efficiency for flexure.

In slabs and walls, it is usually necessary to provide two-way reinforcing. Even where the bending actions occur in only one

direction, the code requires a minimum amount of reinforcing in the other direction for control of cracking due to shrinkage and temperature changes. As shown in Fig. 5-3*a*, even with minimum cover and small bars, a minimum slab thickness is approximately 2 in. Except for joist or waffle construction, however, slab thicknesses are usually greater, for reasons of development of practical levels of flexural resistance. Thus reinforcing is more often as shown in Fig. 5-3*b*, with the bars closer to the top or bottom, depending on whether the moment is positive or negative.

In many cases it is desirable for the slab to have a significant fire rating. Building codes often require additional cover for this purpose, and typically specify minimum slab thicknesses of 4 in. or more. Slab thicknesses required for this purpose also depend on the type of aggregate that is used for the concrete.

Walls of 10 in. or greater thickness often have two separate layers of reinforcing, as shown in Fig. 5-3*c*. Each layer is placed as close as the requirements for cover permit to the outside wall surface. Walls with crisscrossed reinforcing (both vertical and horizontal bars) are seldom made less than 6 in. thick.

As shown in Fig. 5-4, concrete beams usually have a minimum of two reinforcing bars and a stirrup or tie of at least No. 2 or No. 3 size. Even with small bars, the minimum beam width in this situation is at least 8 in., with 10 in. being much more practical.

For rectangular columns with ties, a limit of 8 in. is usual for one wide of an oblong cross section and 10 in. for a square section. Round columns may be either tied or spiral wrapped. A 10-in. diameter may be possible for a round tied column, but 12 in. is more practical and is the usual minimum for a spiral column, with larger sizes required where more cover is necessary.

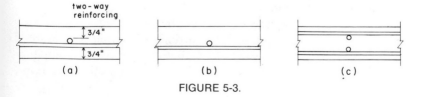

FIGURE 5-3.

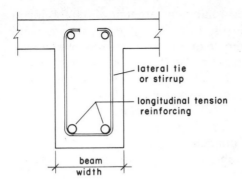

lateral tie
or stirrup

longitudinal tension
reinforcing

beam
width

FIGURE 5-4.

5-6. Shrinkage and Temperature Reinforcement

The essential purpose of steel reinforcing is to prevent the cracking of the concrete due to tension stresses. In the design of concrete structures, investigation is made for the anticipated structural actions that will produce tensile stress: primarily the actions of bending, shear, and torsion. However, tension can also be induced by the shrinkage of the concrete during its drying out after the initial pour. Temperature variations may also induce tension in various situations. To provide for these latter actions, the ACI Code requires a minimum amount of reinforcing in members such as walls and slabs even when structural actions do not indicate any need. These requirements are discussed in the sections that deal with the design of these members.

5-7. Minimum Reinforcing

In the design of most reinforced concrete members the amount of steel reinforcing required is determined from computations and represents the amount determined to be necessary to resist the required tensile force in the member. In various situations, however, there is a minimum amount of reinforcing that is desirable, which may on occasion exceed that determined by the computations. The ACI Code makes provisions for such minimum rein-

forcing in columns, beams, slabs, and walls. The minimum reinforcing may be specified as a minimum percentage of the member cross-sectional area, as a minimum number of bars, or as a minimum bar size. These requirements are discussed in the sections that deal with the design of the various types of members.

6

Flexural Behavior of Beams

||||||||||||||||||||||||||||||||||||||||||||||

6-1. Internal Resisting Moment

In Chapter 4 we learned that bending moment is a measure of the tendency of the external forces on a beam to deform it by bending. We now consider the action within the beam that resists bending, called the resisting moment.

Figure 6-1a represents a simple beam loaded with two concentrated loads. The reactions at the left and right ends are, respectively, R_1 and R_2. In accordance with the laws of equilibrium the algebraic sum of the vertical forces equals zero, the algebraic sum of the horizontal forces equals zero, and the algebraic sum of the moments of the forces equals zero. These three laws are expressed in this manner, $\Sigma\ V = 0$, $\Sigma\ H = 0$, and $\Sigma\ M = 0$. For the present we are concerned only with the law relating to moments.

The resisting moment in a reinforced concrete beam is created by the development of internal stresses which may be represented as a resultant tension T and a resultant compression C, acting on a section $X–X$ taken through the beam (Fig. 6-1b). These two forces constitute a resisting mechanical couple that tends to rotate the beam at section $X–X$ in a counterclockwise direction,

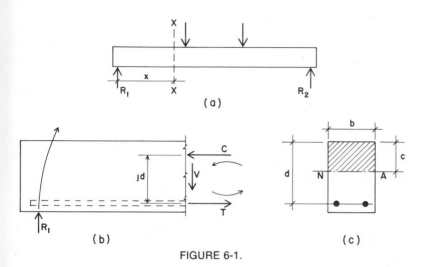

FIGURE 6-1.

thereby opposing the tendency of the bending moment $R_1 \times x$ to cause clockwise rotation. Under equilibrium conditions the resisting moment is always equal to the bending moment. Consequently, if we wish to design a beam for a given loading condition, we arrange its concrete dimensions and the steel reinforcement so that it is capable of developing a resisting moment equal to the maximum bending moment caused by the loads.

Because the reinforcing steel is assumed to carry all the tension, T is located at the centroid of the reinforcement. The force C, however, is the resultant of compressive stresses in the concrete distributed over some portion of the beam depth (c in Fig. 6-1c), which will be established later. The *effective depth* of the beam is the distance from the top (compression) face to the centroid of the steel and is denoted by the symbol d. The additional depth below the reinforcement is not considered in calculations; it provides fire and moisture protection for the steel and assists in developing bond between the concrete and steel. The distance between C and T is called the *arm* of the resisting couple; it is represented by the symbol jd and is discussed further in Section 6-4. The shear force V indicated in Fig. 6-1b is considered in Chapter 7.

6-2. Distribution of Compressive Stress

We know from structural mechanics that when a simple beam is subjected to flexure, as indicated in Fig. 6-1, the upper portion of the member is in compression and the lower portion in tension. Furthermore, there is a horizontal plane separating the compressive and tensile stresses known as the *neutral surface;* at this plane the value of the bending stress is zero. The line in which the neutral surface intersects the beam cross section is called the *neutral axis* (*NA* in Fig. 6-1c) and its distance from the top of the beam is denoted by c.

Deferring consideration of the way in which the value of c is determined, let us direct our attention to Fig. 6-2, which shows three different assumed distributions of stress on the compression side of the neutral axis. The diagram at (*a*) illustrates straight-line distribution in accordance with elastic theory. The stress varies directly as the distance from the neutral axis, at which it is zero, and increases to a maximum value at the compression face of the beam. This value is called the *extreme fiber stress* and is denoted by the symbol f_c. Figure 6-2b illustrates a parabolic distribution when the value of the extreme fiber stress has reached f'_c, the *specified compressive strength* (ultimate strength) of the concrete. This corresponds to inelastic behavior of concrete with an assumed parabolic stress-strain diagram. Figure 6-2c shows a rectangular compressive stress distribution which is assumed to be equivalent in its static effect to the parabolic pattern. The rectan-

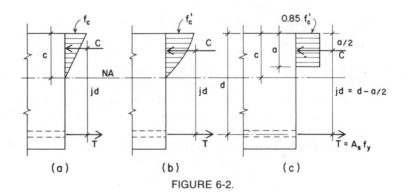

FIGURE 6-2.

gular "stress block" is based on the assumption that a concrete stress of $0.85 f_c'$ is uniformly distributed over a part of the compression zone with dimensions equal to the beam width b and the distance a which locates a line parallel to and above the neutral axis. This is the stress distribution most often used in design under the ultimate strength theory.

6-3. Design Methods and Codes

The design of reinforced concrete structural members may be accomplished by two different methods. One, called *working stress design* (sometimes designated WSD), is based on the straight-line distribution of compressive stress in the concrete (Fig. 6-2a); the other is known as *ultimate strength design* (USD) and is now the predominant design method used for important building structures.

In working stress design a maximum allowable (working) value for the extreme fiber stress is established (Table 6-1) and the formulas are predicated on elastic behavior of the reinforced concrete member under service load. The straight-line distribution of compressive stress is valid at working stress levels because the stresses developed vary approximately with the distance from the neutral axis, in accordance with elastic theory. Shrinkage and cracking of the concrete, however, together with the phenome-

TABLE 6-1. Design Values for Concrete Beams: Working Stress Method[a]

	For values of f_c' in psi shown below:			
	2000	2500	3000	4000
$f_c = 0.45 f_c'$ (psi)	900	1125	1350	1800
Modular ratio: $n = E_x/E_c$	11.3	10.1	9.2	8.0

[a] Adapted from data in *Building Code Requirements for Reinforced Concrete* (ACI-77) with permission of the publisher, American Concrete Institute. Values are for normal weight concrete, 145 lb/ft^3.

non of creep under sustained loading (Section 2-11), complicate the stress distribution. Over time, stresses computed in reinforced concrete members on the basis of elastic theory are not realistic. Generally speaking, the acceptable safety of the working stress design method is maintained by the differentials provided between the allowable compressive stress f_c and the specified compressive strength of the concrete f_c' and between the allowable tensile stress f_s and the yield strength of the steel reinforcement f_y. These, in effect, are measures of safety.

The 1963 Code of the American Concrete Institute contained separate sections that covered working stress design and ultimate strength design, but the 1977 Code, promulgated as *Building Code Requirements for Reinforced Concrete* (ACI 318-77), is built primarily around ultimate strength design. Among reasons contributing to the displacement of WSD in favor of USD is the greater uniformity in safety factors for beams and columns with different loading conditions that is yielded by the latter method. The 1977 ACI Code uses the designation *strength design method* for USD.

The 1977 Code also permits an *alternate design method* that is similar to many of the working stress design procedures of the 1963 Code. For ordinary beams and girders flexural computations are identical with those for WSD under the 1963 Code. There are significant differences, however, in other areas such as design for shear, anchorage length of reinforcement, and design of columns.

The ACI Code is extensive and covers many aspects of ultimate strength design. Only a limited number of its provisions, however, can be discussed in a book of this nature. The treatment herein focuses on design procedures for the common structural elements that occur frequently in building construction, and many items referred to in the Code must necessarily be omitted.

6-4. Flexure Formulas: Working Stress Design

The following is a presentation of the formulas and procedures used in the working stress method. The discussion is limited to a rectangular beam section with tension reinforcing only.

Referring to Figure 6-3, the following are defined:

b = the width of the concrete compression zone

d = the effective depth of the section for stress analysis; from the centroid of the steel to the edge of the compression zone

A_s = the cross-sectional area of the reinforcing

p = the percentage of reinforcing, defined as

$$p = \frac{A_s}{bd}$$

n = the elastic ratio = $\dfrac{E \text{ of the steel reinforcing}}{E \text{ of the concrete}}$

kd = the height of the compression stress zone; used to locate the neutral axis of the stressed section; expressed as a percentage (k) of d

jd = the internal moment arm, between the net tension force and the net compression force; expressed as a percentage (j) of d

f_c = the maximum compressive stress in the concrete

f_s = the tensile stress in the reinforcing

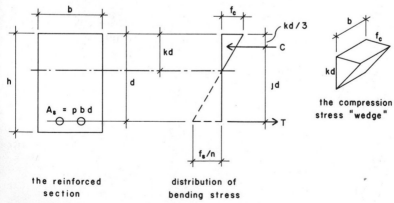

FIGURE 6-3.   Moment resistance of a rectangular concrete section with tension reinforcing.

The compression force C may be expressed as the volume of the compression stress "wedge," as shown in the figure:

$$C = \tfrac{1}{2}(kd)(b)(f_c) = \tfrac{1}{2}kf_cbd$$

Using the compression force, the moment resistance of the section may be expressed as

$$M = Cjd = (\tfrac{1}{2}kf_cbd)(jd) = \tfrac{1}{2}kjf_cbd^2 \tag{1}$$

This may be used to derive an expression for the concrete stress:

$$f_c = \frac{2M}{kjbd^2} \tag{2}$$

The resisting moment may also be expressed in terms of the steel and the steel stress as

$$M = Tjd = (A_s)(f_s)(jd)$$

This may be used for determination of the steel stress or for finding the required area of steel.

$$f_s = \frac{M}{A_s jd} \tag{3}$$

$$A_s = \frac{M}{f_s jd} \tag{4}$$

A useful reference is the so-called balanced section, which occurs when the exact amount of reinforcing used results in the simultaneous limiting stresses in the concrete and steel. The properties which establish this relationship may be expressed as follows:

$$\text{balanced } k = \frac{1}{1 + f_s/nf_c} \tag{5}$$

$$j = 1 - \frac{k}{3} \tag{6}$$

$$p = \frac{f_c k}{2f_s} \tag{7}$$

$$M = Rbd^2 \tag{8}$$

in which

$$R = \tfrac{1}{2}kjf_c \tag{9}$$

[derived from Formula (1)]. If the limiting compression stress in the concrete ($f_c = 0.45\,f'_c$) and the limiting stress in the steel are entered in Formula (5), the balanced section value for k may be found. Then the corresponding values for j, p, and R may be found. The balanced p may be used to determine the maximum amount of tensile reinforcing that may be used in a section without the addition of compressive reinforcing. If less tensile reinforcing is used, the moment will be limited by the steel stress, the maximum stress in the concrete will be below the limit of $0.45\,f'_c$, the value of k will be slightly lower than the balanced value and the value of j slightly higher than the balanced value. These relationships are useful in design for the determination of approximate requirements for cross sections.

Table 6-2 gives the balanced properties for various combinations of concrete strength and limiting steel stress. The values of n, k, j, and p are all without units. However, R must be expressed in particular units; the unit used in the table is kip-inches (k-in.).

TABLE 6-2.   Balanced Section Properties for Rectangular Concrete Sections with Tension Reinforcing Only (Working Stress)

f_s		f'_c		n	k	j	p	R	
ksi	MPa	ksi	MPa					k-in.	kN-m
16	110	2.0	13.79	11.3	0.389	0.870	0.0109	0.152	1045
		2.5	17.24	10.1	0.415	0.862	0.0146	0.201	1382
		3.0	20.68	9.2	0.437	0.854	0.0184	0.252	1733
		4.0	27.58	8.0	0.474	0.842	0.0266	0.359	2468
20	138	2.0	13.79	11.3	0.337	0.888	0.0076	0.135	928
		2.5	17.24	10.1	0.362	0.879	0.0102	0.179	1231
		3.0	20.68	9.2	0.383	0.872	0.0129	0.226	1554
		4.0	27.58	8.0	0.419	0.860	0.0188	0.324	2228
24	165	2.0	13.79	11.3	0.298	0.901	0.0056	0.121	832
		2.5	17.24	10.1	0.321	0.893	0.0075	0.161	1107
		3.0	20.68	9.2	0.341	0.886	0.0096	0.204	1403
		4.0	27.58	8.0	0.375	0.875	0.0141	0.295	2028

When the area of steel used is less than the balanced p, the true value of k may be determined by the following formula:

For $K <$ balanced p

$$k = \sqrt{2np - (np)^2} - np \tag{10}$$

Figure 6-4 may be used to find approximate k values for various combinations of p and n.

6-5. Use of Working Stress Formulas

In the design of concrete beams, there are two situations that commonly occur. The first occurs when the beam is entirely un-

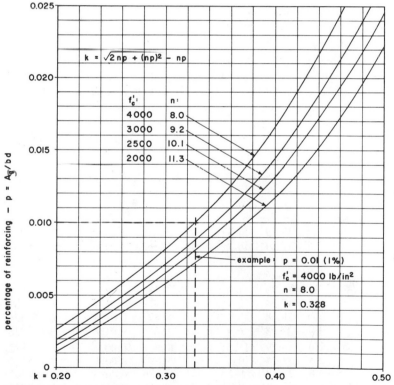

FIGURE 6-4. k factors for rectangular sections with tension reinforcing—as a function of p and n.

determined; that is, the concrete dimensions and the reinforcing are unknown. The second occurs when the concrete dimensions are given, and the required reinforcing for a specific bending moment must be determined. The following examples illustrate the use of the formulas just developed for each of these problems.

Example 1.   A rectangular concrete beam of concrete with f'_c of 3000 psi [20.7 MPa] and steel reinforcing with f_s = 20 ksi [138 MPa] must sustain a bending moment of 200 k-ft [271 kN-m]. Select the beam dimensions and the reinforcing for a section with tension reinforcing only.
Solution: (1) With tension reinforcing only, the minimum size beam will be a balanced section, since a smaller beam would have to be stressed beyond the capacity of the concrete to develop the required moment. Using Formula (8),

$$M = Rbd^2 = 200 \text{ kip-ft [271 kN-m]}$$

Then from Table 6-2, for f'_c of 3000 psi and f_s of 20 ksi,

$$R = 0.226 \text{ (in units of k-in.) [1554 in units of kN-m]}$$

Therefore,

$$M = 200 \times 12 = 0.226(bd^2), \text{ and } bd^2 = 10,619$$
$$[M = 271 = 1554(bd^2), \text{ and } bd^2 = 0.1744]$$

(2) Various combinations of b and d may be found; for example,

$$b = 10 \text{ in.,} \quad d = \sqrt{\frac{10,619}{10}}$$
$$= 32.6 \text{ in., } [b = 0.254 \text{ m, } d = 0.829 \text{ m}]$$
$$b = 15 \text{ in.,} \quad d = \sqrt{\frac{10,619}{15}}$$
$$= 26.6 \text{ in., } [b = 0.381 \text{ m, } d = 0.677 \text{ m}]$$

Although they are not given in this example, there are often some considerations other than flexural behavior alone that influence the choice of specific dimensions for a beam. If the beam is of the ordinary form shown in Fig. 6-5, the specified dimension is

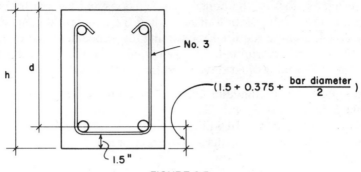

FIGURE 6-5.

usually that given as h. Assuming the use of a No. 3 U-stirrup, a cover of 1.5 in. [38 mm], and an average-size reinforcing bar of 1-in. [25-mm] diameter (No. 8 bar), the design dimension d will be less than h by 2.375 in. [60 mm]. Lacking other considerations, we will assume a b of 15 in. [380 mm] and an h of 29 in. [740 mm], with the resulting d of $29 - 2.375 = 26.625$ in. [680 mm].

(3) We next use the specific value for d with Formula (4) to find the required area of steel A_s. Since our selection is very close to the balanced section, we may use the value of j from Table 6-2. Thus

$$A_s = \frac{M}{f_s j d} = \frac{200 \times 12}{20 \times 0.872 \times 26.625} = 5.17 \text{ in.}^2$$

$$\left[A_s = \frac{271{,}000}{0.138 \times 0.872 \times 680} = 3312 \text{ mm}^2 \right]$$

Or using the formula for the definition of p and the balanced p value from Table 6-2,

$$A_s = pbd = 0.0129(15 \times 26.625) = 5.15 \text{ in.}^2$$

$$[A_s = 0.0129(380 \times 680) = 3333 \text{ mm}^2]$$

(4) We next select a set of reinforcing bars to obtain this area. As with the beam dimensions, there are other concerns. For the

purpose of our example, if we select bars all of a single size (see the table on inside of book cover), the number required will be:

For No. 6 bars, $\dfrac{5.17}{0.44} = 11.75,$ or 12 $\left[\dfrac{3312}{284} = 11.66\right]$

For No. 7 bars, $\dfrac{5.17}{0.60} = 8.62,$ or 9 $\left[\dfrac{3312}{387} = 8.56\right]$

For No. 8 bars, $\dfrac{5.17}{0.79} = 6.54,$ or 7 $\left[\dfrac{3312}{510} = 6.49\right]$

For No. 9 bars, $\dfrac{5.17}{1.00} = 5.17,$ or 6 $\left[\dfrac{3312}{645} = 5.13\right]$

For No. 10 bars, $\dfrac{5.17}{1.27} = 4.07,$ or 5 $\left[\dfrac{3312}{819} = 4.04\right]$

For No. 11 bars, $\dfrac{5.17}{1.56} = 3.31,$ or 4 $\left[\dfrac{3312}{1006} = 3.29\right]$

For all except the No. 11 bars, the requirements for bar spacing (as discussed in Section 5-3) would result in the need to place the bars in stacked layers in the 15-in.-wide beam. While this is possible, it would require some increase in the dimension h in order to maintain the effective depth of approximately 26.6 in., since the centroid of the steel bar areas would move farther away from the edge of the concrete.

Example 2. A rectangular concrete beam of concrete with f'_c of 3000 psi [20.7 MPa] and steel with f_s of 20 ksi [138 MPa] has dimensions of $b = 15$ in. [380 mm] and $h = 36$ in. [910 mm]. Find the area required for the steel reinforcing for a moment of 200 kip-ft [271 kN-m].
Solution: The first step in this case is to determine the balanced moment capacity of the beam with the given dimensions. If we assume the section to be as shown in Fig. 6-5, we may assume an approximate value for d to be h minus 2.5 in. [64 mm], or 33.5 in. [851 mm]. Then with the value for R from Table 6-2,

$$M = Rbd^2 = 0.226 \times 15 \times (33.5)^2 = 3804 \text{ k-in.}$$

or $M = \dfrac{3804}{12} = 317$ k-ft

$$[M = 1554 \times 0.380 \times (0.850)^2 = 427 \text{ kN-m}]$$

Since this value is considerably larger than the required moment, it is thus established that the given section is larger than that required for a balanced stress condition. As a result, the concrete flexural stress will be lower than the limit of $0.45 f'_c$, and the section is qualified as being under-reinforced; which is to say that the reinforcing required will be less than that required to produce a balanced section (with moment capacity of 317 k-ft). In order to find the required area of steel, we use Formula (4) just as we did in the preceding example. However, the true value for j in the formula will be something greater than that for the balanced section (0.872 from Table 6-2).

As the amount of reinforcing in the section decreases below the full amount required for a balanced section, the value of k decreases and the value of j increases. However, the range for j is small: from 0.872 up to something less than 1.0. A reasonable procedure is to assume a value for j, find the corresponding required area, and then perform an investigation to verify the assumed value for j, as follows.

Assume $j = 0.90$

Then

$$A_s = \frac{M}{f_s jd} = \frac{200 \times 12}{20 \times 0.90 \times 33.5} = 3.98 \text{ in.}^2$$

and

$$p = \frac{A_s}{bd} = \frac{3.98}{15 \times 33.5} = 0.00792$$

$$\left[A_s = \frac{271{,}000}{0.138 \times 0.90 \times 850} = 2567 \text{ mm}^2 \right.$$

$$\left. p = \frac{2567}{380 \times 850} = 0.00795 \right]$$

Using this value for p in Fig. 6-4, we find $k = 0.313$. Using Formula (6), we then determine j to be

$$j = 1 - \frac{k}{3} = 1 - \frac{0.313}{3} = 0.896$$

which is reasonably close to our assumption, so the computed area is adequate for design.

Problem 6-5-A.   A rectangular concrete beam has concrete with $f'_c = 3000$ psi [20.7 MPa] and steel reinforcing with $f_s = 20$ ksi [138 MPa]. Select the beam dimensions and reinforcing for a balanced section if the beam sustains a bending moment of 240 k-ft [325 kN-m].

Problem 6-5-B.   Find the area of steel reinforcing required and select the bars for the beam in Problem 6-5-A if the section dimensions are $b = 16$ in. and $d = 32$ in.

6-6. Flexure Formulas: Strength Design

Figure 6-6 shows the equivalent rectangular compressive stress distribution in the concrete permitted by the ACI Code for use in the strength design method. As stated in Section 6-2, the rectangular stress block is based on the assumption that, at ultimate load, a concrete stress of $0.85\,f'_c$ is uniformly distributed over the compression zone. The dimensions of this zone are the beam width b and the distance a which locates a line parallel to and above the neutral axis. Although we have not yet considered how

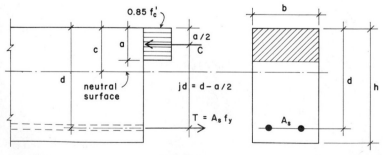

FIGURE 6-6.

the value of a is determined, let us turn our attention to Fig. 6-6 and develop equations for the theoretical resisting moment M_t.

We observe that the resultant (sum) of the compressive stresses is

$$C = 0.85 f'_c \times b \times a$$

and that it acts at a distance of $a/2$ from the top of the beam. The arm of the resisting moment couple jd then becomes $d - a/2$ and the theoretical resisting moment as governed by the concrete is

$$M_t = C \left(d - \frac{a}{2} \right) = (0.85 f'_c ba) \times \left(d - \frac{a}{2} \right) \qquad (1)$$

Similarly, the theoretical moment strength as controlled by the steel reinforcement is

$$M_t = T \left(d - \frac{a}{2} \right) = (A_s f_y) \left(d - \frac{a}{2} \right) \qquad (2)$$

If *balanced* conditions exist, that is, if the concrete reaches its full compressive strength when the steel reaches its yield strength, the two equations will be equal to each other, or

$$0.85 f'_c ba = A_s f_y = \rho bd f_y \qquad (3)$$

where

$$\rho = \frac{A_s}{bd}$$

Note: The ACI Code uses ρ to indicate steel percent with the strength method, whereas p is used with the working stress method. Then, from Formula (3)

$$a = \frac{\rho bd f_y}{0.85 f'_c b} = \frac{\rho f_y d}{0.85 f'_c} \qquad (4)$$

and

$$\rho = \frac{a}{d} \times \frac{0.85 f'_c}{f_y} \qquad (5)$$

The symbol ρ_b is used to denote the balancing ratio of reinforcement and a_b, the depth of the stress block under balanced conditions.

Using this expression for a and considering the strain relationship between concrete and steel, the following formula for the balancing ratio of reinforcement may be derived:

$$\rho_b = \frac{0.85 f'_c \beta_1}{f_y} \times \frac{87,000}{87,000 + f_y} \tag{6}$$

in which β_1 is a coefficient relating the depth of the rectangular stress block to the depth from the compression face to the neutral axis, or $a = \beta_1 \times c$. (See Fig. 6-6). The value of β_1 varies with the strength of the concrete. The ACI Code prescribes a value of 0.85 for concrete strengths up to 4000 psi [27.6 MPa] and a reduction of 0.05 for each 1000 psi [6.895 MPa] of strength in excess of 4000 psi, with a minimum value of 0.65. (For example, if $f'_c = 5000$ psi [34.5 MPa], $\beta_1 = 0.80$, and so on.)

By equating Formulas (5) and (6), we can derive an expression for the balanced value of a/d. Thus

$$\frac{a}{d} \times \frac{0.85 f'_c}{f_y} = \frac{0.85 f'_c \beta_1}{f_y} \times \frac{87,000}{87,000 + f_y}$$

and

$$\frac{a}{d} = \beta_1 \frac{87,000}{87,000 + f_y} \tag{7}$$

$$\left[\begin{array}{l} \text{For SI units:} \\[2mm] \dfrac{a}{d} = \beta_1 \dfrac{600,000}{600,000 + f_y} \\[2mm] \text{when stress is in MPa} \end{array} \right]$$

Referring to Formula (1), we may derive another form for this equation as follows:

$$M_t = 0.85 f'_c ba \left(d - \frac{a}{2} \right)$$

$$= 0.85 f'_c b \frac{a}{d} d^2 \left(1 - \frac{1}{2} \frac{a}{d} \right)$$

$$= bd^2 \left[0.85 f'_c \left\{ \frac{a}{d} - \frac{1}{2} \left(\frac{a}{d} \right)^2 \right\} \right]$$

$$= R bd^2 \tag{8}$$

where

$$R = 0.85 f'_c \left[\frac{a}{d} - \frac{1}{2} \left(\frac{a}{d} \right)^2 \right] \tag{9}$$

Formulas (5), (7), and (9) may be used to derive factors for balanced sections which can be used in the design of beams. Table 6-3 contains a compilation of these factors for five values of concrete strength and two values of steel yield strength. The use of this material is demonstrated in the following section.

6-7.   Use of Strength Design Formulas

Use of the strength design formulas derived in Section 6.6 is illustrated in the following examples.

Example 1.   The service load bending moments on a rectangular beam 10 in. [254 mm] wide are 58 k-ft [78.6 kN-m] for dead load and 38 k-ft [51.5 kN-m] for live load. If f'_c = 4000 psi [27.6 MPa] and f_y = 60 ksi [414 MPa], determine the depth of the beam and the required area of tension reinforcing.

Solution: (1) The required ultimate moment strength (M_u) is first determined in accordance with Section 4-9.

$$U = 1.4D + 1.7L$$

$$M_u = 1.4(M_{DL}) + 1.7(M_{LL})$$

$$= 1.4(58) + 1.7(38) = 146 \text{ k-ft}$$

$$[M_u = 1.4(78.6) + 1.7(51.5) = 198 \text{ kN-m}]$$

(2) To find the required design moment strength (M_t) we apply the capacity reduction factor ϕ = 0.90 (Section 4-9) and the relationship $M_u = \phi M_t$; thus

$$M_t = \frac{M_u}{\phi} = \frac{146}{0.90} = 162 \text{ k-ft} \quad \text{or} \quad 1944 \text{ k-in.}$$

$$\left[M_t = \frac{198}{0.90} = 220 \text{ kN-m} \right]$$

(3) The maximum usable reinforcement ratio as given in Table

TABLE 6-3. Balanced Section Properties for Rectangular Concrete Sections with Tension Reinforcing Only: Strength Design[a]

| f'_c | | $f_y = 40$ ksi [276 MPa] | | | | | $f_y = 60$ ksi [414 MPa] | | | | |
psi	MPa	Balanced a/d	Usable a/d (75% Balance)	Usable ρ	Usable R k-in.	kN-m	Balanced a/d	Usable a/d (75% Balance)	Usable ρ	Usable R k-in.	kN-m
2000	13.79	0.5823	0.4367	0.0186	0.580	4000	0.5031	0.3773	0.0107	0.520	3600
2500	17.24	0.5823	0.4367	0.0232	0.725	5000	0.5031	0.3773	0.0137	0.650	4500
3000	20.69	0.5823	0.4367	0.0278	0.870	6000	0.5031	0.3773	0.0160	0.781	5400
4000	27.58	0.5823	0.4367	0.0371	1.161	8000	0.5031	0.3773	0.0214	1.041	7200
5000	34.48	0.5480	0.4110	0.0437	1.388	9600	0.4735	0.3551	0.0252	1.241	8600

[a] See Section 6-6 for derivation of formulas used to obtain table values.

69

6-3 is $\rho = 0.0214$. If a balanced section is used, we may thus determine the required area of reinforcement from the relationship

$$A_s = \rho b d$$

While there is nothing especially desirable about a balanced section, it does represent the beam section with least depth if tension reinforcing only is used. We will therefore proceed to find the required balanced section for this example.

(4) To determine the required effective depth d, we use Formula (8) from Section 6-6; thus

$$M_t = R b d^2$$

With the value of $R = 1.041$ from Table 6-3,

$$M_t = 1944 = 1.041(10)(d)^2$$

and

$$d = \sqrt{\frac{1944}{1.041(10)}} = \sqrt{186.7} = 13.66 \text{ in.}$$

$$\left[d = \sqrt{\frac{220}{7200(0.254)}} = 0.347 \text{ m} \right]$$

(5) If this value is used for d, the required steel area may be found as

$$A_s = \rho b d = 0.0214 \times 10 \times 13.66 = 2.92 \text{ in.}^2$$

The ACI Code requires a minimum ratio of reinforcing as follows:

$$A_{s\min} = \frac{200}{f_y} = \frac{200}{60,000} = 0.0033$$

which is clearly not critical for this example.

Selection of the actual beam dimensions and the actual number and size of reinforcing bars would involve various considerations, as discussed in Section 6-5 and illustrated in Fig. 6-5. We will not complete the example in this case, since more complete design situations will be illustrated in the examples in the succeeding chapters.

If there are reasons, as there often are, for not selecting the least deep section with the greatest amount of reinforcing, a slightly different procedure must be used, as illustrated in the following example.

Example 2.   Using the same data as in Example 1, find the reinforcing required if the desired beam section has $b = 10$ in. [254 mm] and $d = 18$ in. [457 mm].

Solution: The first two steps in this situation would be the same as in Example 1—to determine M_u and M_t. The next step would be to determine whether the given section is larger than, smaller than, or equal to a balanced section. Since this investigation has already been done in Example 1, we may observe that the 10 in. by 18 in. section is larger than a balanced section. Thus the actual value of a/d will be less than the balanced section value of 0.3773. The next step would then be as follows:

(4) Estimate a value for a/d—something smaller than the balanced value. For example, try $a/d = 0.25$. Then

$$a = 0.25d = 0.25(18) = 4.5 \text{ in. [114 mm]}$$

With this assumed value for a, we may use Formula (2) to find a value for A_s.

(5) Referring to Section 6-6 and Fig. 6-6,

$$A_s = \frac{M_t}{f_y(d - a/2)} = \frac{1944}{60(15.75)} = 2.057 \text{ in.}^2$$

(6) We next test to see if the estimate for a/d was close by finding a/d using Formula (5) of Section 6-6. Thus:

$$\rho = \frac{A_s}{bd} = \frac{2.057}{10 \times 18} = 0.0114$$

and

$$\frac{a}{d} = \rho \frac{f_y}{0.85 f_c'} = 0.0114 \frac{60}{0.85(4)} = 0.202$$

$$a = 0.202(18) = 3.63 \text{ in.}, \quad d - \frac{a}{2} = 16.2 \text{ in.}$$

If we replace the value for $d = a/2$ that was used earlier with this new value, the required value of A_s will be slightly reduced. In this example, the correction will be only a few percent. If the first guess for a/d had been way off, it may justify a second run through steps 4, 5, and 6 to get closer to an exact answer.

Problem 6-7-A. A rectangular beam has concrete with $f_c' = 3000$ psi [20.7 MPa] and steel reinforcing with $f_y = 40$ ksi [276 MPa]. Using strength methods, find the depth required and the area of steel required for a balanced section if the beam width is 16 in. [406 mm]. Service load dead load moment is 140 k-ft [190 kN-m] and live load moment is 100 k-ft [136 kN-m].

Problem 6-7-B. Find the steel area required for the section in Problem 6-7-A if the section dimensions are $b = 16$ in. [406 mm] and $d = 32$ in. [812 mm].

(Compare the answers obtained in these problems with those obtained in Problems 6-5-A and 6-5-B, in which the problem data is similar, but the work is done by the working stress method.)

7

Shear and Diagonal Tension

IIIIIIIIIIIIIIIIIIIIIIIIIIIIIIIIIIIIIIIIIIII

Shear is developed in a number of situations in concrete structures. In most cases, it is not the resulting shear stresses them-
selves that are of primary concern, but rather the *diagonal tension* stresses that ordinarily accompany shear stresses. The
material in this chapter deals with the general problem of shear
with a concentration on shear in beams. Investigation and design
for shear in slabs, walls, and footings is further discussed in other
chapters.

7-1. Shear Situations in Concrete Structures

The most common situations involving shear in concrete structures are shown in Fig. 7-1. Shear in beams (Fig. 7-1*a*) is ordinarily critical near the supports, where the shear force is greatest. In
short brackets (Fig. 7-1*b*) and keys (Fig. 7-1*c*) the shear action is
essentially a direct slicing effect. Punching shear, also called peripheral shear, occurs in column footings and in slabs that are
directly supported on columns (Fig. 7-1*d*). When walls are used
as bracing elements for shear forces that are parallel to the wall

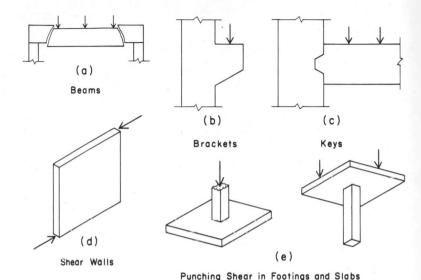

(a)

Beams

(b)

Brackets

(c)

Keys

(d)

Shear Walls

(e)

Punching Shear in Footings and Slabs

FIGURE 7-1.   Situations involving shear in concrete structures.

surface (called shear walls), they must develop resistance to the direct shear effect that is similar to that in a bracket.

In all of these situations consideration must be given to the shear effect and the resulting shear stresses. Both the magnitude and the direction of the shear stresses must be considered. In many cases, however, the shear effect occurs in combination with other effects, such as bending moment, axial tension, or axial compression. In combined force situations the resulting net combined stress situations must be considered.

7-2.   Development of Shear Stress

Shear force generates a lateral, slicing effect in materials. Visualized in two dimensions, this direct effect is as shown in Fig. 7-2a. For stability within the material, there will be a counteracting, or reactive shear stress developed at right angles to the active stress, as shown in Fig. 7-2b. Finally, the interaction of the direct and reactive shears will produce both diagonal tension and diagonal

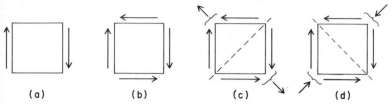

FIGURE 7-2.   Development of shear stress.

compression stresses, as shown in Fig. 7-2c and 7-2d. In concrete the critical stress concern is for the diagonal tension stress, which works on the weakest property of the material.

Referring to Fig. 7-2, we may observe that:

1. The unit reactive (right angle) shear stress will be equal in magnitude to the unit active shear stress.
2. The combined diagonal effect (tension or compression) will be $\sqrt{2}$ times the shear effect (vector combination of the direct and reactive shears).
3. The diagonal stress will develop on the diagonal plane, which is $\sqrt{2}$ times the size of the plane on which the shear stress is developed. Thus the diagonal stress will be the same magnitude as the shear stress; both the force effect and the stressed area being $\sqrt{2}$ times that for shear.

Accepting the observations just made, it is possible to determine the critical tension stress that accompanies shear by simply computing the unit shear stress. As in other situations, however, it is also necessary to determine the *direction* of the tension stress, especially when reinforcement must be provided.

The stress actions shown in Fig. 7-2 represent the conditions that occur when shear alone is considered. When shear occurs simultaneously with other effects, the various resulting stress conditions must be combined to produce the net effect. Figure 7-3 shows the result of combining a shear stress effect with a direct tension stress effect. For shear alone, the critical tension stress plane is at 45°, as shown in Fig. 7-2a. For the tension alone, the critical tension stress plane is at 90°. For the combined stress condition, the unit stress will be some magnitude higher than

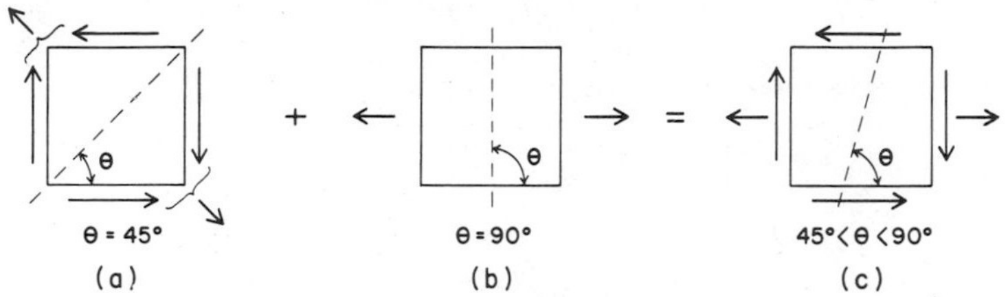

$\theta = 45°$ $\theta = 90°$ $45° < \theta < 90°$
(a)   (b)   (c)

FIGURE 7-3.   Development of combined stress.

cither the shear or direct tension stress, and the critical tension stress plane will be at an angle somewhere between 45 and 90°.

Although most of our investigations for shear will consider the shear effects alone, the full effects of any combined stress conditions must also be investigated for a complete evaluation of behaviors.

7-3. Shear in Beams

Let us consider the case of a simple beam with uniformly distributed load and end supports that provide only vertical resistance (no moment restraint). The distribution of internal shear and bending moment are as shown in Fig. 4-3b. For flexural resistance, it is necessary to provide longitudinal reinforcing bars near the bottom of the beam. These bars are oriented for primary effectiveness in resistance to tension stresses that develop on a vertical (90°) plane (which is the case at the center of the span, where the bending moment is maximum and the shear approaches zero).

Under the combined effects of shear and bending, the beam tends to develop tension cracks as shown in Fig. 7-4a. Near the center of the span, where the bending is predominant and the shear approaches zero, these cracks approach 90°. Near the support, however, where the shear predominates and bending approaches zero, the critical tension stress plane approaches 45°, and the horizontal bars are only partly effective in resisting the cracking.

For beams, the most common form of shear reinforcement

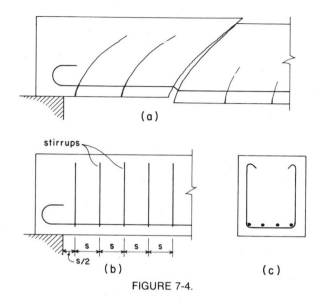

FIGURE 7-4.

consists of a series of U-shaped bent bars (Fig. 7-4c), placed vertically and spaced along the beam span, as shown in Fig. 7-4b. These bars are intended to provide a vertical component of resistance, working in conjunction with the horizontal resistance provided by the flexural reinforcement. In order to develop tension near the support face, the horizontal bars must be bonded to the concrete beyond the point where the stress is developed. Where the beam ends extend only a short distance over the support (a common situation), it is often necessary to bend or hook the bars, as shown in Fig. 7-4.

The simple span beam and the rectangular section shown in Fig. 7-4 occur only infrequently in building structures. The most common case is that of the beam section shown in Fig. 7-5a, which occurs when a beam is poured monolithically with a supported concrete slab. In addition, these beams normally occur in continuous spans with negative moments at the supports. Thus the stress in the beam near the support is as shown in Fig. 7-5a, with the negative moment producing compressive flexural stress in the bottom of the beam stem. This is substantially different

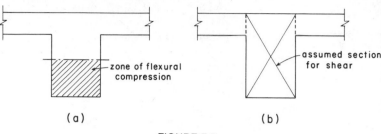

FIGURE 7-5.

from the case of the simple beam, where the moment approaches zero near the support.

For purpose of shear resistance, the continuous, T-shaped beam is considered to consist of the section indicated in Fig. 7-5b. The effect of the slab is ignored, and the section is considered to be a simple rectangular one. Thus for shear design, there is little difference between the simple span beam and the continuous beam, except for the effect of the continuity on the distribution of shear along the beam span. It is important, however, to understand the relationships between shear and moment in the continuous beam.

Figure 7-6 illustrates the typical condition for an interior span of a continuous beam with uniformly distributed load. Referring to the portions of the beam span numbered 1, 2, and 3, we note:

1.   In this zone the high negative moment requires major flexural reinforcing consisting of horizontal bars near the top of the beam.
2.   In this zone, the moment reverses sign; moment magnitudes are low; and, if shear stress is high, the design for shear is a predominant concern.
3.   In this zone, shear consideration is minor and the predominant concern is for positive moment requiring major flexural reinforcing in the bottom of the beam.

Vertical U-shaped stirrups, similar to those shown in Fig. 7-7a, may be used in the T-shaped beam. An alternate detail for the U-shaped stirrup is shown in Fig. 7-7b, in which the top hooks are turned outward; this makes it possible to spread the negative

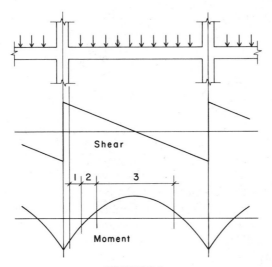

FIGURE 7-6.

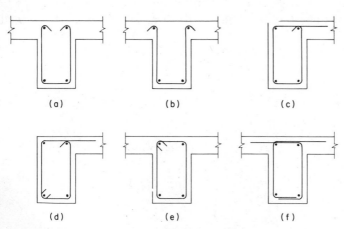

(a)   (b)   (c)

(d)   (e)   (f)

FIGURE 7-7.   Various forms for vertical stirrups.

moment reinforcing bars to make placing of the concrete some-what easier. Figures 7-7c and 7-7d show possibilities for stirrups in beams that occur at the edges of large openings or at the out-side edge of the structure. This form of stirrup is used to enhance the torsional resistance of the section and also assists in develop-ing the negative moment resistance in the slab at the edge of the beam.

So-called closed stirrups, similar to ties in columns, are some-times used for T-shaped beams, as shown in Figs. 7-7e and 7-7f. These are generally used to improve the torsional resistance of the beam section.

Stirrup forms are often modified by designers or by the rein-forcing fabricator's detailers to simplify the fabrication and/or the field installation. The stirrups shown in Figures 7-7d and 7-7f are two such modifications of the basic details in Figs. 7-7c and 7-7e, respectively.

7-4. Beam Shear: General Design Considerations

The following are some of the general considerations and code requirements that apply to current practices of design for beam shear.

Concrete Capacity.   Whereas the tensile strength of the con-crete is ignored in design for flexure, the concrete is assumed to take some portion of the shear in beams. If the capacity of the concrete is not exceeded—as it sometimes is for lightly loaded beams—there may be no need for reinforcing. The typical case, however, is as shown in Fig. 7-8, where the maximum shear V exceeds the capacity of the concrete alone (V_c) and the steel

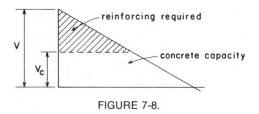

FIGURE 7-8.

reinforcing is required to absorb the excess, indicated as the shaded portion in the shear diagram.

Minimum Shear Reinforcing.   Even when the maximum computed shear stress falls below the capacity of the concrete, the present code requires the use of some minimum amount of shear reinforcing. Exceptions are made in some situations, such as for slabs and very shallow beams. The objective is essentially to toughen the structure with a small investment in additional reinforcing.

Type of Stirrup.   The most common stirrups are the simple U shape or closed forms shown in Fig. 7-7, placed in a vertical position at intervals along the beam. It is also possible to place stirrups at an incline (usually 45°), which makes them somewhat more effective in direct resistance to the potential shear cracking near the beam ends. (See Fig. 7-4.) In large beams with excessively high unit shear stress, both vertical and inclined stirrups are sometimes used at the location of the greatest shear.

Size of Stirrups.   For beams of moderate size, the most common size for U-stirrups is a No. 3 bar. These bars can be bent relatively tightly at the corners (small radius of bend) in order to fit within the beam section. For larger beams, a No. 4 bar is sometimes used, its strength (as a function of its cross-sectional area) being almost twice that of a No. 3 bar.

Reinforcing for Narrow Beams.   When beams are less than about 10 in. wide, it is not possible to bend a U-shaped stirrup to fit within the beam profile. If shear reinforcing is required, one form that is used is the so-called *ladder* stirrup, shown in Fig. 7-9.

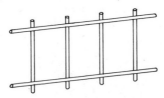

FIGURE 7-9.   "Ladder" shear reinforcing for a narrow beam.

This consists of a series of single vertical bars welded to horizontal bars at the top and bottom. A variation on this type of reinforcing consists of using a portion of heavy-gage welded wire fabric.

Spacing of Stirrups.   Stirrup spacings are computed (as discussed in the following sections) on the basis of the amount of reinforcing required for the unit shear stress at the location of the stirrups. A maximum spacing of $d/2$ (i.e., one-half the effective beam depth d) is specified in order to assure that at least one stirrup occurs at the location of any potential diagonal crack. (See Fig. 7-4.) When shear stress is excessive, the maximum spacing is limited to $d/4$.

Critical Maximum Design Shear.   Although the actual maximum shear value occurs at the end of the beam, the code permits the use of the shear stress at a distance of d (effective beam depth) from the beam end as the critical maximum for stirrup design. Thus, as shown in Fig. 7-10, the shear requiring reinforcing is slightly different from that shown in Fig. 7-8.

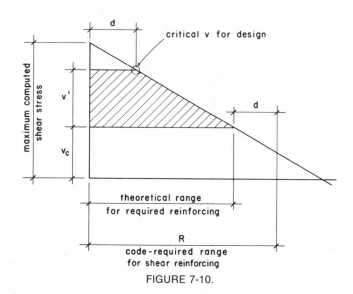

FIGURE 7-10.

Total Length for Shear Reinforcing. On the basis of computed shear stresses, reinforcing must be provided along the beam length for the distance defined by the shaded portion of the shear stress diagram shown in Fig. 7-10. For the center portion of the span, the concrete is theoretically capable of the necessary shear resistance without the assistance of reinforcing. However, the code requires that some reinforcing be provided for a distance beyond this computed cutoff point. The 1963 ACI Code required that stirrups be provided for a distance equal to the effective depth of the beam beyond the cutoff point. The 1977 ACI Code requires that minimum shear reinforcing be provided as long as the computed shear stress exceeds one-half of the capacity of the concrete. However it is established, the total extended range over which reinforcing must be provided is indicated as R on Fig. 7-10.

7-5. Design of Shear Reinforcement: Working Stress Method

The following is a description of the procedure for design of shear reinforcing for beams that is in compliance with Appendix B of the 1977 ACI Code (Ref. 1).

Shear stress is computed as

$$v = \frac{V}{bd}$$

in which V = the total shear force at the section
b = the beam width (of the stem for T-shapes)
d = the effective depth of the section

For beams of normal weight concrete, subjected only to flexure and shear, shear stress in the concrete is limited to

$$v_c = 1.1 \sqrt{f_c'}$$

When v exceeds the limit for v_c, reinforcing must be provided, complying with the general requirements discussed in Section 7-4. Although the code does not use the term, we coin the nota-

tion of v' for the excess unit shear for which reinforcing is required. Thus

$$v' = v - v_c$$

Required spacing of shear reinforcement is determined as follows. Referring to Fig. 7-11, we note that the capacity in tensile resistance of a single, two-legged stirrup is equal to the product of the total steel cross-sectional area times the allowable steel stress. Thus

$$T = (A_v) \times (f_s)$$

This resisting force opposes the development of shear stress on the area s times b, where b is the width of the beam and s is the spacing (half the distance to the next stirrup on each side). Equating the stirrup tension to this force, we obtain the equilibrium equation

$$(A_v) \times (f_s) = (b) \times (s) \times (v')$$

From this equation, we can derive an expression for the required spacing; thus

$$s = \frac{A_v f_s}{v' b}$$

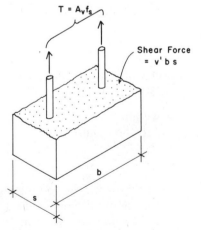

FIGURE 7-11.

The following example illustrates the procedure for a simple beam.

Example 1.   Using the working stress method, design the required shear reinforcing for the simple beam shown in Fig. 7-12. Use $f'_c = 3$ ksi [20.7 MPa] and $f_s = 20$ ksi [138 MPa] and single U-shaped stirrups.
Solution: The shear diagram for the beam will be of the form

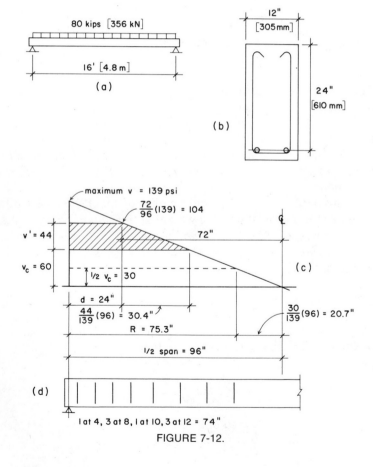

FIGURE 7-12.

shown in Fig. 4-3*b*. The maximum value for the shear is 40 k [178 kN] and the maximum value for shear stress is computed as

$$v = \frac{V}{bd} = \frac{40,000}{12 \times 24} = 139 \text{ psi } [957 \text{ kPa}]$$

We now construct the shear stress diagram for one half of the beam, as shown in Fig. 7-12*c*. For the shear design, we determine the critical shear stress at 24 in. (the effective depth of the beam) from the support. Using proportionate triangles, this value is

$$\frac{72}{96}(139) = 104 \text{ psi } [718 \text{ kPa}]$$

The capacity of the concrete without reinforcing is

$$v_c = 1.1 \sqrt{f_c'} = 1.1 \sqrt{3000} = 60 \text{ psi } [414 \text{ kPa}]$$

At the point of critical stress, therefore, there is an excess shear stress of $104 - 60 = 44$ psi $[718 - 414 = 304$ kPa] that must be carried by reinforcing. We next complete the construction of the diagram in Fig. 7-12*c* to define the shaded portion, which indicates the extent of the required reinforcing. We thus observe that the excess shear condition extends to 54.4 in. [1.382 m] from the support.

In order to satisfy the requirements of the 1977 ACI Code, shear reinforcing must be used wherever the computed unit stress exceeds one-half of v_c. As shown in Fig. 7-12*c*, this is a distance of 75.3 in. from the support. The code further stipulates that the minimum cross-sectional area of this reinforcing be

$$A_v = 50 \frac{bs}{f_y}$$

If we assume an f_y value of 50 ksi [345 MPa] and use the maximum allowable spacing of one-half the effective depth, the required area is

$$A_v = 50 \frac{12 \times 12}{50,000} = 0.144 \text{ in.}^2$$

which is less than the area of $2 \times 0.11 = 0.22$ in.^{2} provided by the two legs of the No. 3 stirrup.

For the maximum v' value of 44 ksi, the maximum spacing required is determined as

$$s = \frac{A_v f_s}{v'b} = \frac{0.22 \text{ in.}^2 \times 20,000 \text{ psi}}{44 \text{ psi} \times 12 \text{ in.}} = 8.3 \text{ in.}$$

Since this is less than the maximum allowable of 12 in., it is best to calculate at least one more spacing at a short distance beyond the critical point. We thus determine that the unit stress at 36 in. from the support is

$$v = \frac{60}{96} \times 139 = 87 \text{ psi}$$

and the value of v' at this point is $87 - 60 = 27$ psi. The spacing required at this point is thus

$$s = \frac{0.22 \times 20,000}{27 \times 12} = 13.6 \text{ in.}$$

which indicates that the required spacing drops to the maximum allowed at less than 12 in. from the critical point. A possible choice for the stirrup spacings is shown in Fig. 7-12d, with a total of eight stirrups that extend over a range of 74 in. from the support. There are thus a total of 16 stirrups in the beam, 8 at each end.

Example 2. Determine the required number and spacings for No. 3 U-stirrups for the beam shown in Fig. 7-13. Use $f'_c = 3$ ksi [20.7 MPa] and $f_s = 20$ ksi [138 MPa].
Solution: As in the preceding example, the shear values and corresponding stresses are determined, and the diagram in Fig. 7-13c is constructed. In this case, the maximum critical shear stress of 89 psi results in a maximum v' value of 29 psi, for which the required spacing is

$$s = \frac{0.22 \times 20,000}{29 \times 10} = 15.2 \text{ in.}$$

Since this value exceeds the maximum limit of $d/2 = 10$ in., the stirrups may all be placed at the limiting spacing, and a possible arrangement is as shown in Fig. 7-13d.

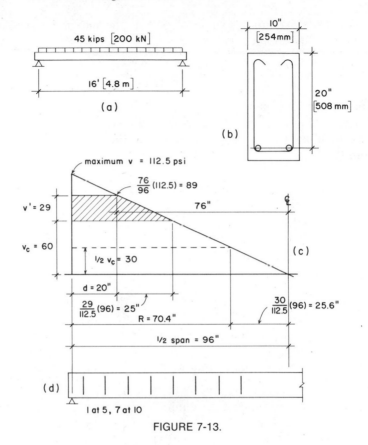

FIGURE 7-13.

Note that in both examples the first stirrup is placed at one-half the required distance from the support.

Example 3.   Determine the required number and spacings for No. 3 U-stirrups for the beam shown in Fig. 7-14. Use $f'_c = 3$ ksi [20.7 MPa] and $f_s = 20$ ksi [138 MPa].

Solution: In this case, the maximum critical design shear stress is found to be less than v_c, which in theory indicates that reinforcing is not required. To comply with the code requirement for minimum reinforcing, however, we provide stirrups at the maximum permitted spacing out to the point where the shear stress

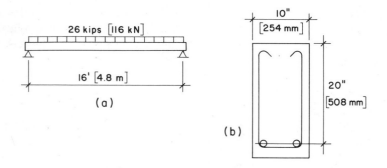

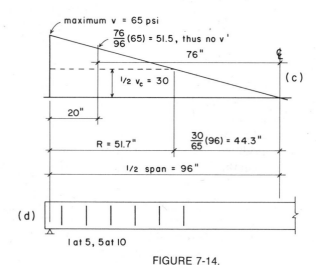

FIGURE 7-14.

drops to 30 psi (one-half of v_c). To verify that the No. 3 stirrup is adequate, we compute

$$A_v = 50 \, \frac{10 \times 10}{50,000} = 0.10 \text{ in.}^2 \quad \text{(See Example 1.)}$$

which is less than the area of 0.22 in. provided, so the No. 3 stirrup at 10-in. spacing is adequate.

The preceding examples have illustrated what is generally the simplest case for beam shear design—that of a beam with uni-

formly distributed load and with sections subjected only to flexure and shear. When concentrated loads or unsymmetrical loadings produce other forms for the shear diagram, these must be used for design of the shear reinforcing. In addition, where axial forces of tension or compression exist in the concrete frame, consideration must be given to the combined effects when designing for shear.

When torional moments exist (twisting moments at right angles to the beam), their effects must be combined with beam shear.

Problem 7-5-A.   A concrete beam similar to that shown in Fig. 7-13 sustains a total load of 60 kips [267 kN] on a span of 24 ft [7.32 m]. Determine the layout for a set of No. 3 U-stirrups using $f_s = 20$ ksi [138 MPa] and $f'_c = 3000$ psi [20.7 MPa]. The section dimensions are $b = 12$ in. [305 mm] and $d = 26$ in. [660 mm].

7-6. Design of Shear Reinforcement: Strength Design

The requirements and procedures for strength design are essentially similar to those for working stress design. The principal difference is in the use of ultimate resistance as opposed to working stresses at service loads. The basic requirement in strength design is that the modified ultimate resistance of the section be equal to or greater than the factored load. This condition is stated as

$$V_u \leq \phi V_n$$

in which V_u = the factored shear force at the section
V_n = the nominal shear strength of the section

The nominal strength is defined as

$$V_n = V_c + V_s$$

in which V_c is nominal strength provided by concrete
V_s is nominal strength provided by reinforcing

The term *nominal strength* is used to differentiate between the computed resistances and the usable value of total resistance, which is reduced for design by the strength reduction factor ϕ. (See Section 4.9 for a general discussion of strength design.)

For members subjected to shear and flexure only, the nominal concrete strength is defined as

$$V_c = 2 \sqrt{f'_c} \, bd$$

Translated into unit stress terms, this means that the limiting nominal shear stress in the concrete is $2 \sqrt{f'_c}$, and when reduced by ϕ, the limiting *working* ultimate strength is $0.85 \times 2 \sqrt{f'_c} = 1.7 \sqrt{f'_c}$.

When shear reinforcing consists of vertical stirrups, the nominal reinforcing strength is defined as

$$V_s = \frac{A_v f_y d}{s}$$

with a limiting value for V_s established as

$$V_s = 8 \sqrt{f'_c} \, bd$$

The following example illustrates the use of strength design methods for shear reinforcing. The problem data is essentially the same as for Example 1 in Section 7-5, so that a comparison of the design results can be made.

Example.   Using strength design methods, determine the spacing required for No. 3 U-stirrups for the beam shown in Fig. 7-15. Use $f'_c = 3$ ksi [20.7 MPa] and $f_y = 50$ ksi [345 MPa].

Solution: The loads shown in Fig. 7-15a are service loads. These must be converted to *factored loads* for strength design, as discussed in Section 4-9. We thus determine the factored load to be

$$W_u = 1.4(\text{dead load}) + 1.7(\text{live load})$$

$$= 1.4(40) + 1.7(40)$$

$$= 124 \text{ k}$$

The maximum shear force is thus 62 k, and the shear diagram for one-half the beam is as shown in Fig. 7-15c. The critical value for V_u at 24 in. (effective beam depth) from the support is deter-

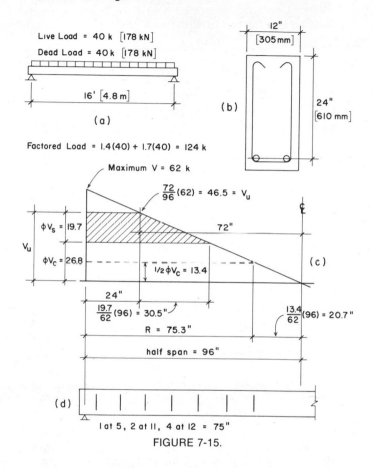

FIGURE 7-15.

mined from proportionate triangles to be 46.5 k. The usable capacity of the concrete is determined as

$$\phi V_c = \phi 2 \sqrt{f_c'} \, bd$$
$$= 0.85(2 \sqrt{3000})(12 \times 24)$$
$$= 26{,}816 \text{ lb or approximately } 26.8 \text{ k}$$

and for the reinforcing

$$\phi V_s = 46.5 - 26.8 = 19.7 \text{ k}$$

Therefore,

$$V_s = \frac{19.7}{\phi} = \frac{19.7}{0.85} = 23.18 \text{ k}$$

and the required spacing is determined from

$$V_s = \frac{A_v f_y d}{s}$$

$$s = \frac{A_v f_y d}{V_s} = \frac{0.22 \times 50 \times 24}{23.18} = 11.4 \text{ in.}$$

Referring to Example 1 in Section 7-5, we may see that this value is larger than that computed by the working stress method; thus the strength design is somewhat less conservative for this example.

A possible choice of stirrup spacings is that shown in Fig. 7-15d, using seven stirrups at each end of the beam.

To verify that the value for V_s is within the limit previously given, we compute the maximum value of

$$V_s = 8 \sqrt{f_c'} \, bd = 8 \sqrt{3000} \times 12 \times 24 = 126 \text{ k}$$

which is far from critical.

Problem 7-6-A. Redo Problem 7-5-A using strength design methods.

8

Development
of Reinforcement
||||||||||||||||||||||||||||||||||||||||||||

8-1. Bond Stress

Bond stress is the essential interactive relationship between a
steel reinforcing bar and the concrete mass of the structural element in which it is embedded. Bond stresses are developed on the
surfaces of all reinforcing bars whenever some structural action
requires the steel and concrete to share load. The basic concept
of bond stress development may be illustrated by the simple example shown in Fig. 8-1, in which a steel bar is embedded in a
block of concrete and is required to resist a pull-out tension force.

Figure 8-1b shows the static equilibrium relationship for the
steel bar, with the pull-out force developed as the product of a
tensile stress times the area of the bar cross section ($f_s \times \pi D^2/4$)
and the resisting force developed by a bond stress (u) operating
on the surface of the bar ($u \times \pi D \times L$). By equating these two
forces, we can derive an expression either for the unit bond stress
or the required embedment length for a limiting bond stress.

Bond stress development is affected by a number of considerations; some of the major ones are the following:

1. Grade of Steel.   As the f_y of the steel is increased, the
    allowable f_s value will also increase, requiring the develop-

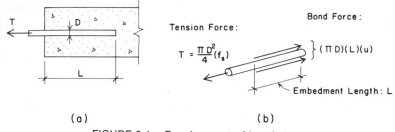

Tension Force:

$$T = \frac{\pi D^2}{4}(f_s)$$

Bond Force:

$\left.\right\}$ (π D)(L)(u)

Embedment Length: L

(a)   (b)

FIGURE 8-1. Development of bond stress.

ment of higher bond stresses or the need for greater embedment lengths.

2. Strength of Concrete.   In general, as f'_c is increased, the capability for development of bond stress is also increased.

3. Bar Size.   Consideration of the expression for the tension force in the bar in Fig. 8-1 will indicate that the force capability of the bar increases with the square of the diameter. On the other hand, the resistance developed by bond stress increases only linearly with increase of the bar diameter. Thus bond stresses tend to be more critical on bars of large diameter.

4. Concrete Encasement.   The bonding force must be developed in the concrete mass around each bar. This development is limited when this mass is constrained due to closely spaced groups of bars or where bars are placed close to the edge of the concrete member.

5. Location of Bars.   When concrete is poured into forms and cured into its hardened state, the concrete near the bottom of the member tends to develop slightly higher quality than that near the top. The weight of the concrete mass above produces a denser material in the lower concrete, and the exposed top surface tends to dry more rapidly, resulting in less well-cured concrete near the top. This difference in quality affects the potential for bond resistance, so some adjustment is made for bars placed near the top (such as reinforcement for negative moment in beams).

In times past, working stress procedures included the establishment of allowable stresses for bond and the computation of bond stresses for various situations. At present, however, the codes deal with this problem as one of development length, as discussed in the next section.

8-2. Development of Reinforcement

The 1977 ACI Code defines *development length* as the length of embedment required to develop the design strength of the reinforcing at a critical section. For beams, critical sections occur at points of maximum stress and at points within the span where some of the reinforcement terminates or is bent up or down. For a uniformly loaded simple span beam, one critical section is at midspan, where the bending moment is a maximum. The tensile reinforcing required for flexure at this point must extend on both sides a sufficient distance to develop the stress in the bars; however, except for very short spans with large bars, the bar lengths will ordinarily be more than sufficient.

In the simple beam, the bottom reinforcing required for the maximum moment at midspan is not entirely required as the moment decreases toward the end of the span. It is thus sometimes the practice to make only part of the midspan reinforcing continuous for the whole beam length. In this case it may be necessary to assure that the bars that are of partial length are extended sufficiently from the midspan point and that the bars remaining beyond the cutoff point can develop the stress required at that point.

When beams are continuous through the supports, top reinforcing is required for the negative moments at the supports. These top bars must be investigated for the development lengths in terms of the distance they extend from the supports.

8-3. Development of Tensile Reinforcement

For tension reinforcing consisting of bars of No. 11 size and smaller, the code specifies a minimum length for development (l_d) as follows:

$$l_d = 0.04A_b \frac{f_y}{\sqrt{f'_c}}$$

but not less than $0.0004d_b f_y$ or 12 in.
In these formulas A_b is the cross-sectional area of the bar and d_b is the bar diameter.

Modification factors for l_d are given for various situations, as follows:

For top bars in horizontal members with at least
12 in. of concrete below the bars   1.4

For sets of bars where the bars are 6 in. or more
on center   0.8

For flexural reinforcement that is in excess of
that required by computations $\dfrac{A_s \text{ required}}{A_s \text{ provided}}$

Additional modification factors are given for lightweight concrete, for bars encased in spirals, and for bars with f_y in excess of 60 ksi.

Table 8-1 gives values for minimum development lengths for tensile reinforcing, based on the requirements of the 1977 ACI Code. The values listed under Other Bars are the unmodified length requirements; those listed under Top Bars are increased by the modification factor for this situation. Values are given for two concrete strengths and for the two most commonly used grades of tensile reinforcing.

8-4.   Use of Hooked Ends

When details of the construction restrict the ability to extend bars sufficiently to produce required development lengths, partial development can sometimes be achieved by use of a hooked end. Section 12.5 of the 1977 ACI Code provides a means by which a so-called standard hook may be evaluated in terms of an equivalent development length. Detailed requirements for standard hooks are given in Chapter 7 of the 1977 ACI Code. Bar ends may

TABLE 8-1. Minimum Development Length for Tensile Reinforcement (in.)

| Bar size (no.) | $f_y = 40$ ksi [276 MPa] | | | | $f_y = 60$ ksi [414 MPa] | | | |
| | $f'_c = 3$ ksi [20.7 MPa] | | $f'_c = 4$ ksi [27.6 MPa] | | $f'_c = 3$ ksi [20.7 MPa] | | $f'_c = 4$ ksi [27.6 MPa] | |
	Top bars[a]	Other bars	Top bars[a]	Other bars	Top bars[a]	Other bars	Top bars[a]	Other bars
3	12	12	12	12	13	12	13	12
4	12	12	12	12	17	12	17	12
5	14	12	12	12	21	15	21	15
6	18	13	16	12	27	19	15	18
7	25	18	21	15	37	26	32	23
8	32	23	28	20	48	35	42	30
9	41	29	36	25	61	44	53	38
10	52	37	45	32	78	56	68	48
11	64	46	55	40	96	68	83	59
14	87	62	75	54	130	93	113	81
18	113	80	98	70	169	120	146	104

Note: Lengths are based on requirements of the 1977 ACI Code.
[a] Horizontal bars so placed that more than 12 in. [305 mm] of concrete is cast in the member below the reinforcement.

be bent at 90, 135, or 180° to produce a hook. The 135° bend is used only for ties and stirrups, which normally consist of relatively small diameter bars. (See Fig. 5-2.)

Table 8-2 gives values for standard hooks, using the same variables for f'_c and f_y that are used in Table 8-1. The table values given are in terms of the equivalent development length provided by the hook. Comparison of the values in Table 8-2 with those given for the unmodified lengths (Other) in Table 8-1, will show that the hooks are mostly capable of only partial development. The development length provided by a hook may be added to whatever development length is provided by extension of the bar, so that the total development may provide for full utilization of the bar tension capacity (at f_y) in many cases. The following example illustrates the use of the data from Tables 8-1 and 8-2 for a simple situation.

TABLE 8-2. Equivalent Embedment Lengths of Standard Hooks (in.)

Bar size	$f_y = 40$ ksi [276 MPa]		$f_y = 60$ ksi [414 MPa]	
	$f'_c = 3$ ksi [20.7 MPa]	$f'_c = 4$ ksi [27.6 MPa]	$f'_c = 3$ ksi [20.7 MPa]	$f'_c = 4$ ksi [27.6 MPa]
3	3.0	3.4	4.4	5.1
4	3.9	4.5	5.9	6.8
5	4.9	5.7	7.4	8.5
6	6.3	6.8	9.5	10.2
7	8.6	8.6	12.9	12.9
8	11.4	11.4	17.1	17.1
9	14.4	14.4	21.6	21.6
10	18.3	18.3	24.4	24.4
11	22.5	22.5	26.2	26.2

Example. The negative moment in the short cantilever shown in Fig. 8-2 is resisted by the steel bar in the top of the beam. Determine whether the development of the reinforcing is adequate. Use $f'_c = 3$ ksi [20.7 MPa] and $f_y = 60$ ksi [414 MPa].

Solution: The maximum moment in the cantilever is produced at the face of the support; thus the full tensile capacity of the bar should be developed on both sides of this section. In the beam itself the condition is assumed to be that of a "top bar," for which Table 8-1 yields a required minimum development length of 27 in., indicating that the length of 46 in. provided is more than adequate. Within the support, the condition is unmodified, and the requirement is for a length of 19 in. The actual extended development length provided within the support is 14 in., which is measured as the distance to the end of the hooked bar end, as shown in the figure. If the hooked end qualifies as a *standard hook* (in accordance with the requirements of Chapter 7 of the 1977 ACI Code), the equivalent development length provided (from Table 8-2) is 9.5 in. Thus the total development provided by the combination of extension and hooking is 14 + 9.5 = 23.5 in., which exceeds the requirement of 19 in., so the development is adequate.

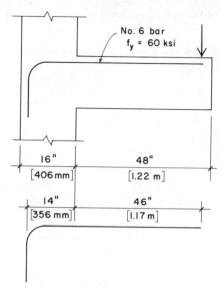

FIGURE 8-2.

In a real situation, it is probably not necessary to achieve the full development lengths given in Table 8-1, since bar selection often results in some slight excess in the actual steel cross-sectional area provided. In such a case, the required development length can be reduced by the modification factor given in Section 8-3.

Problems 8-4-A to 8-4-D. Determine whether the bar shown in Fig. 8-3 is adequately anchored for development of the full f_y capacity of the steel.

	f'_c		f_y		L_2		L_1		Bar size
	ksi	MPa	ksi	MPa	in.	mm.	in.	mm.	
A	3	20.7	60	414	36	914	12	305	8
B	4	27.6	60	414	42	1067	16	206	10
C	3	20.7	40	276	38	965	18	457	9
D	4	27.6	40	276	35	889	17	432	9

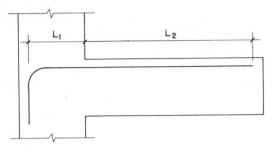

FIGURE 8-3.

8-5. Development of Compressive Reinforcement

The discussion of development length so far has dealt with tension bars only. Development length in compression is, of course, a factor in column design and in the design of beams reinforced for compression.

The absence of flexural tension cracks in the portions of beams where compression reinforcement is employed, plus the beneficial effect of the end bearing of the bars on the concrete, permit shorter development lengths in compression than in tension. The ACI Code prescribes that l_d for bars in compression shall be computed by the formula

$$l_d = \frac{0.02 f_y d_b}{\sqrt{f'_c}}$$

but shall not be less than $0.0003 f_y d_b$ or 8 in., whichever is greater. Table 8-3 lists compression bar development lengths for a few combinations of specification data.

8-6. Bar Development in Simple Beams

The ACI Code defines *development length* as the length of embedded reinforcement required to develop the design strength of the reinforcement at a critical section. Critical sections occur at points of maximum stress and at points within the span at which adjacent reinforcement terminates or is bent up into the top of the

TABLE 8-3.   Minimum Development Length for Compressive Reinforcement (in.)

Bar size	$f_y = 40$ ksi [276 MPa]		$f_y = 60$ ksi [414 MPa]	
	$f'_c = 3$ ksi [20.7 MPa]	$f'_c = 4$ ksi [27.6 MPa]	$f'_c = 3$ ksi [20.7 MPa]	$f'_c = 4$ ksi [27.6 MPa]
3	8.0	8.0	8.0	8.0
4	8.0	8.0	11.0	9.5
5	9.2	8.0	13.7	11.9
6	10.9	9,5	16.4	14.2
7	12.8	11.1	19.2	16.6
8	14.6	12.7	21.9	19.0
9	16.5	14.3	24.8	21.5
10	18.5	16.1	27.8	24.1
11	20.6	17.9	31.0	26.8
14	—	—	37.1	32.1
18	—	—	49.5	42.8

beam. For a uniformly loaded simple beam, one critical section is at midspan where the bending moment is maximum. This is a point of maximum tensile stress in the reinforcement (peak bar stress) and some length of bar is required over which the stress can be developed. Other critical sections occur between midspan and the reactions at points where some bars are cut off because they are no longer needed to resist the bending moment; such terminations create peak stress in the remaining bars that extend the full length of the beam.

Example 1.   A uniformly loaded rectangular beam is used on a simple span of 16 ft. It sustains a theoretical bending moment of $M_t = 1944$ k-in., induced by service dead and live loads of 29 and 19 k, respectively; these produce a factored uniform load of $W_u = 73.0$ k. The beam is 10 in. wide and has an effective depth of 14 in. The tension reinforcement consists of four No. 8 bars, all of which extend the entire length of the beam. Compute the required development length of the reinforcement if $f_y = 60,000$ psi and $f'_c = 4000$ psi.

Solution: Referring to Table 1-2, the area of an individual No. 8 bar is found to be 0.79 in.[2]. Then

$$l_d = \frac{0.04 A_b f_y}{\sqrt{f_c'}} = \frac{0.04 \times 0.79 \times 60,000}{\sqrt{4000}} = 30 \text{ in.}$$

Because all bars extend the full length of the beam, an embedment length (from section of peak bar stress at midspan) equal to 8 ft, or 96 in., is provided. This greatly exceeds the required development length of 30 in., and consequently l_d is not a critical factor in this beam.

Example 2. If two of the four bars in the beam of Example 1 are cut off short of each end at sections beyond which they are no longer needed for bending moment, compute the required development length of the other two bars that continue into the supports.

Solution: (1) The parabolic curve in Fig. 8-4 represents the bending moment diagram for this uniformly loaded simple beam. Although presumably there is the usual small discrepancy between A_s required and that actually supplied (due to use of standard bar sizes), we neglect this difference and assume that the potential resisting moment of the four bars just matches the 1944 k-in. developed by the loading.

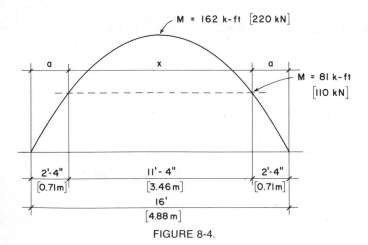

FIGURE 8-4.

(2) The dashed line in the figure indicates the resisting moment of half the total reinforcement. Where this line intersects the bending moment curve two of the four bars are no longer necessary and may be cut off. For a uniformly loaded simple beam this intersection falls between $\frac{1}{6}$ and $\frac{1}{7}$ of the span length from the support. Taking $L/7$ as the approximate position, the distance in this case is $16 \div 7 = 2.29$ ft or 2 ft 4 in. (28 in.).

(3) At this cutoff point the peak stress in the *continuing* bars is the same as at midspan, and consequently a development length of 30 in. (see Example 1 above or Table 8-1) is required beyond this critical section. Because the section occurs 28 in. from the end of the span, l_d is critical by this test and a standard hook must be made at the end of the bars. Referring to Table 8-2, we note that the equivalent embedment length of a standard hook on a bottom No. 8 bar for our specification data is 17.1 in. This is more than adequate to cover the $30 - 28 = 2$ in. of extra length required here.

(4) An additional condition must be satisfied at simple supports. The diameter of the reinforcement must be small enough so that the computed development length of the bar will not exceed $(M_t/V_u) + l_a$. This may be expressed as

$$l_d \lessgtr \frac{M_t}{V_u} + l_a$$

bars. The standard hook must be retained, however, to satisfy the other requirements of Step 3.

(5) Reference to Fig. 8-4 will show that if the *terminated* bars were actually cut off at the points of intersection with the moment curve their overall length would be 11 ft 4 in. However, the ACI Code requires that reinforcement extend beyond the point at which it is no longer needed in bending for a distance equal to the effective depth of the beam or 12 bar diameters, whichever is greater. In our example the effective depth is 14 in., and 12 times the nominal diameter of a No. 8 bar is 12 in. Therefore the actual length of the terminated bars will be 11 ft 4 in., plus twice 14 in., or 13 ft 8 in.

In practice it would have to be decided whether the cost of

fabricating hooks on both ends of the two continuing bars might exceed that of the extra steel used if all four bars continued into the supports (as in Example 1). The simpler fabrication involved and the greater ease of placing the reinforcement in the forms would probably favor selection in this case of four full-length straight bars.

Problem 8-6-A. A simply supported rectangular beam has a width of 10 in., an effective depth of 15 in., and tension reinforcement consisting of four No. 7 bars. It sustains a theoretical bending moment of 1260 k-in., induced by a factored uniform load of $W_u = 54$ k. The span is 14 ft 0 in. center-to-center of bearing areas on masonry walls. Two of the four bars are cut off at the $\frac{1}{4}$ points of the span and the remaining bars continue into the supports, extending 3 in. beyond the center of bearing at each end. The beam has been designed on the basis of $f'_c = 3000$ psi and $f_y = 40,000$ psi. Check the required development length of the continuing bars against that provided. Is it adequate?

Problem 8-6-B. A rectangular concrete beam is used on a simple span of 21 ft 0 in. It is 12 in. wide, has an effective depth of 19 in., and is reinforced with four No. 8 bars. The beam supports a total factored uniform load of $W_u = 66.4$ k which produces a theoretical bending moment of 2320 k-in. Two of the four bars are cut off at the $\frac{1}{4}$ points of the span and the remaining bars continue into the supports, extending 3 in. beyond the center of bearing at each end. The specification data for the design were $f'_c = 4000$ psi and $f_y = 40,000$ psi. Check the required development length of the continuing bars against the embedment length provided. Is it adequate?

8-7. Bar Development in Continuous Beams

When beams are continuous through their supports, the negative moments at the supports will require that bars be placed in the top of the beams. Within the span, bars will be required in the bottom of the beam for the positive moments. While the positive moment will go to zero at some distance from the supports, the codes require that some of the positive moment reinforcing be extended for the full length of the span and a short distance into the support.

Figure 8-5 shows a possible layout for reinforcing in a beam with continuous spans and a cantilevered end at the first support. Referring to the notation in the illustration, we make the following observations.

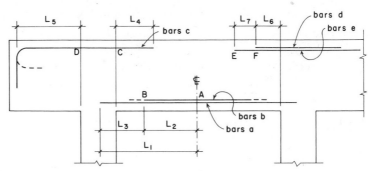

FIGURE 8-5.   Various situations for consideration of development length.

1. Bars a and b are provided for the maximum moment of positive sign that occurs somewhere near the beam midspan. If all these bars are made full length (as shown for bars a), the length L_1 must be sufficient for development (this situation is seldom critical). If bars b are partial length, as shown in the illustration, then length L_2 must be sufficient to develop bars b and length L_3 must be sufficient to develop bars a. As was discussed for the simple beam, the partial length bars must actually extend beyond the theoretical cutoff point (B in the illustration) and the true length must include the dotted portions indicated for bars b.

2. For the bars at the cantilevered end, the distances L_4 and L_5 must be sufficient for development of bars c. L_4 is required to extend beyond the actual cutoff point of the negative moment by the extra length described for the partial length bottom bars. If L_5 is not adequate, the bar ends may be bent into the 90° hook as shown or the 180° hook shown by the dotted line.

3. If the combination of bars shown in the illustration is used at the interior support, L_6 must be adequate for the development of bars d and L_7 adequate for the development of bars e.

For a single loading condition on a continuous beam it is possible to determine specific values of moment and their location along the span, including the locations of points of zero moment. In practice, however, most continuous beams are designed for more than a single loading condition, which further complicates the problems of determining development lengths required.

8-8. Bar Development in Columns

In reinforced concrete columns both the concrete and the steel bars share the compression force. Ordinary construction practices require the consideration of various situations for development of the stress in the reinforcing bars. Figure 8-6 shows a multistory concrete column with its base supported on a concrete footing. With reference to the illustration, we note the following.

1. The concrete construction is ordinarily produced in multiple, separate pours, with construction joints between the separate pours occurring as shown in the illustration.

2. In the lower column, the load from the concrete is transferred to the footing in direct compressive bearing at the joint between the column and footing. The load from the reinforcing must

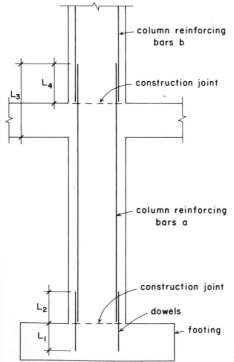

FIGURE 8-6. Development considerations for columns.

be developed by extension of the reinforcing into the footing: distance L_1 in the illustration. Although it may be possible to place the column bars in position during pouring of the footing to achieve this, the common practice is to use dowels, as shown in the illustration. These dowels must be developed on both sides of the joint: L_1 in the footing and L_2 in the column. If the f'_c value for both the footing and the column are the same, these two required lengths will be the same.

3. The lower column will ordinarily be cast together with the supported concrete framing above it, with a construction joint occurring at the top level of the framing (bottom of the upper column), as shown in the illustration. The distance L_3 is that required to develop the reinforcing in the lower column—bars a in the illustration. As for the condition at the top of the footing, the distance L_4 is required to develop the reinforcing in bars b in the upper column. L_4 is more likely to be the critical consideration for the determination of the extension required for bars a.

8-9. Splices

In various situations in reinforced concrete structures it becomes necessary to transfer stress between steel bars in the same direction. Continuity of force in the bars is achieved by splicing, which may be affected by welding, by mechanical means, or by the so-called lapped splice. Figure 8-7 illustrates the concept of the lapped splice, which consists essentially of the development of both bars within the concrete. The length of the lap becomes the development length for both bars. Because a lapped splice is usually made with the two bars in contact, the lapped length must

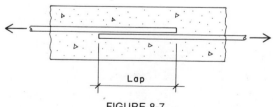

FIGURE 8-7.

usually be somewhat greater than the simple development length required in Table 8-1.

Sections 12-15 to 12.20 of the 1977 ACI Code give requirements for various types of splices. For a simple tension lap splice, the full development of the bars requires a lap length of 1.7 times that required for simple development of the bars. Lap splices are generally limited to bars of No. 11 size and smaller.

For pure tension members, lapped splicing is not permitted, and splicing must be achieved by welding the bars or by some mechanical connection. End-to-end butt welding of bars is usually limited to compression splicing of large diameter bars with high f_y for which lapping is not feasible.

When members have several reinforcing bars that must be spliced, the splicing must be staggered. In general, splicing is not desirable, and is to be avoided where possible. Because bars are obtainable only in limited lengths, however, some situations unavoidably involve splicing. Horizontal reinforcing in long walls is one such case.

9

Design of Beams
||||||||||||||||||||||||||||||||||||||||||||||||||||

9-1. General

Most concrete beams, except in precast structures, occur as parts of monolithic, poured-in-place systems. The various types of common systems are discussed in Chapter 10. The design of a single beam involves a large number of pieces of data most of which are established for the system as a whole, rather than individually for each beam.

Data affecting a single beam, which are usually established for the whole framing system, includes the following.

Concrete Strength f_c'. This is ordinarily chosen as a single value for the whole system or at least for all the parts poured in a single, continuous pour. Ordinary strengths used range from an f_c' of 3 ksi [20.7 MPa] to 5 ksi [34.5 MPa]. For minor structures, a value of 2 ksi [13.8 MPa] is sometimes used, as some local codes permit the omission of batch sampling and testing if this low strength is used for design; this is a major economy factor for small buildings.

Reinforcing Grade f_y. Availability of steel is somewhat a local consideration. The two most often used grades for beams have f_y of 40 or 60 ksi [276 or 414 MPa].

Type of Concrete. Most concrete is made with ordinary sand and gravel. However, the lack of availability of good gravel or the desire for weight reduction may affect the choice of other materials. With regard to beam design, the type of aggregate used will affect the weight, the modulus of elasticity, and long-term creep of the concrete.

System Type and Layout. Beam spans, loading, end conditions, dimensional constraints, and other factors may be determined by the choice of the type of system and the specific form and dimensions of the system plan layout. These considerations are discussed in Chapter 10.

Slab Thickness. Most beams occur in conjunction with solid concrete slabs which are poured monolithically with the beams. Slab thickness is usually established by the structural requirements of the spanning slab, and, in some cases, by consideration of the thermal, acoustic, and fire-resistive properties of the slab.

With all of the foregoing data ordinarily established on a system-wide basis, the design of a single beam is usually limited to the following considerations:

Choice of shape and dimensions of the beam cross section.

Selection of the type and size of shear reinforcing.

Choice of the minimum concrete cover for the beam reinforcing. This is ordinarily done to satisfy the code requirements, based on exposure conditions or fire-resistive rating.

Choice of the flexural reinforcing to satisfy the steel area requirements at the critical sections of the beam.

9-2. Beam Shape

Figure 9-1 shows the most common shapes used for beams in poured-in-place construction. The single, simple rectangular section is actually uncommon, but does occur in some situations. Design of the concrete section consists of selecting the two dimensions: the width b and the overall height or depth h.

As mentioned previously, beams occur most often in conjunc-

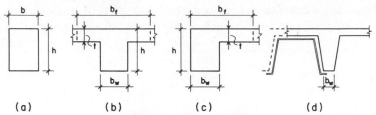

FIGURE 9-1.   Common shapes for concrete beam cross sections.

tion with monolithic slabs, resulting in the typical T shape shown in Fig. 9-1*b*, or the L shape shown in Fig. 9-1*c*. The full T shape occurs at the interior portions of the system, while the L shape occurs at the outside edge of the system or at the side of large openings. As shown in the illustration, there are four basic dimensions for the T and L that must be established in order to fully define the beam section.

t is the slab thickness; it is ordinarily established on its own, rather than as a part of the single beam design.

h is the overall beam stem depth, corresponding to the same dimension for the rectangular section.

b_w is the beam stem width, which is critical for consideration of shear and for problems of fitting reinforcing into the section.

b_f is the so-called *effective width* of the flange, which is the portion of the slab assumed to work with the beam.

A special beam shape is that shown in Fig. 9-1*d*. This occurs in concrete joist and waffle construction when "pans" of steel or reinforced plastic are used to form the concrete, the taper of the beam stem being required for easy removal of the forms. The smallest width dimension of the beam stem is ordinarily used for the beam design in this situation.

9-3.   Beam Width

The width of a beam will affect its resistance to bending. Consideration of the flexure formulas given in Sections 6-4 and 6-6 will show that the width dimension affects the bending resistance in a linear relationship (double the width and you double the resisting

moment, etc.). On the other hand, the resisting moment is affected by the *square* of the effective beam depth. Thus efficiency—in terms of beam weight or concrete volume—will be obtained by striving for deep, narrow beams, instead of shallow, wide ones. (Just as a 2 × 8 is more efficient than a 4 × 4 in wood.)

Beam width also relates to various other factors, however, and these are often critical in establishing the minimum width for a given beam. The formula for shear stress ($v = V/bd$) indicates that the depth is less effective in shear resistance than in moment resistance. Placement of reinforcing bars is also a problem in narrow beams. Table 9-1 gives minimum beam widths required for various bar combinations, based on considerations of bar spacing (Section 5-3), minimum concrete cover of 1.5 in., and use of a No. 3 stirrup. Situations requiring additional concrete cover, use of larger stirrups, or the intersection of beams with columns, may necessitate widths greater than those given in Table 9-1.

9-4. Beam Depth

For specification of the construction, the beam depth is defined by the overall concrete dimension: h in Fig. 9-1. For structural design, however, the critical depth dimension is that from the center of the tension reinforcing to the far side of the concrete: d in Fig. 9-2. While the selection of the depth is partly a matter of

TABLE 9-1.   Minimum Beam Widths[a]

Number of bars	Bar size								
	3	4	5	6	7	8	9	10	11
2	10	10	10	10	10	10	10	10	10
3	10	10	10	10	10	10	10	11	11
4	10	10	10	10	11	11	12	13	14
5	10	11	11	12	12	13	14	16	17
6	11	12	13	14	14	15	17	18	20

[a] Minimum width in inches for beams with 1.5-in. cover, No. 3 U-stirrups, clear spacing between bars of one bar diameter or minimum of 1.0 in. Minimum practical width for beam with No. 3 U-stirrups: 10 in.

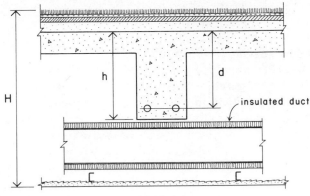

FIGURE 9-2.

satisfying structural requirements, it is typically constrained by other considerations in the building design.

Figure 9-2 shows a section through a typical building floor/ceiling with a concrete beam-slab structure. In this situation the critical depth from a general building design point of view is the overall thickness of the construction, shown as H in the illustration. In addition to the concrete structure, this includes allowances for the floor finish, the ceiling framing, and the passage of an insulated air duct. The net usable portion of H for the structure is shown as the dimension h, with the effective structural depth d being something less than h. Since the space defined by H is not highly usable for the building occupancy, there is a tendency to constrain it which works to limit the extravagant use of d.

Most concrete beams tend to fall within a limited range in terms of the ratio of width to depth. The typical range is for a width/depth ratio between $1:1.5$ and $1:2.5$, with an average of $1:2$. This is not a code requirement or a magic rule; it is merely the result of satisfying typical requirements for flexure, shear, bar spacing, and deflection.

9-5. Deflection Control

Deflection of spanning slabs and beams of poured-in-place concrete is controlled primarily by using recommended minimum

thicknesses (overall height) expressed as a percentage of the span. Table 9-2 is adapted from a similar table given in Section 9.5 of the ACI Code and yields minimum thicknesses as a fraction of the span. Table values apply only for concrete of normal weight (made with ordinary sand and gravel) and for reinforcing with f_y of 60 ksi [414 MPa]. The Code supplies correction factors for other concrete weights and reinforcing grades. The Code further stipulates that these recommendations apply only where beam deflections are not critical for other elements of the building construction, such as supported partitions subject to cracking caused by beam deflections.

Table 9-3 yields maximum spans for beams with various overall depths. These are based on the requirements given in Table 9-2. It should be noted that these are *limits* and are not necessarily practical or efficient values. Use of these limits will usually result in beams having a great amount of reinforcing, whereas economy is generally achieved by using minimum amounts of reinforcing.

TABLE 9-2. Minimum Thickness of One-Way Slabs or Beams Unless Deflections Are Computed

Type of member	End conditions	Minimum thickness of slab or height of beam	
		$f_y = 40$ ksi [276 MPa]	$f_y = 60$ ksi [414 MPa]
Solid one-way slabs[a]	Simple support	$L/25$	$L/20$
	One end continuous	$L/30$	$L/24$
	Both ends continuous	$L/35$	$L/28$
	Cantilever	$L/12.5$	$L/10$
Beams or joists	Simple support	$L/20$	$L/16$
	One end continuous	$L/23$	$L/18.5$
	Both ends continuous	$L/26$	$L/21$
	Cantilever	$L/10$	$L/8$

Source: Data adapted from *Building Code Requirements for Reinforced Concrete* (ACI 318-77), 1977 ed., with permission of the publishers, American Concrete Institute.

[a] Valid only for members not supporting or attached to partitions or other construction likely to be damaged by large deflections.

TABLE 9-3.   Maximum Spans for Beams[a]

Overall beam depth h (in.)	Maximum permissible span (ft)			
	Simply supported	One end continuous	Both ends continuous	Cantilever
10	13.3	15.4	17.5	6.7
12	16	18.5	21	8
14	18.7	21.6	24.5	9.3
16	21.3	24.7	28	10.7
18	24	27.7	31.5	12
20	26.7	30.8	35	13.3
24	32	37.0	42	16
30	40	46.2	52.5	20
36	48	55.5	63	24

[a] Based on requirements of Table 9-2. For normal weight concrete and reinforcing with $f_y = 60$ ksi. For $f_y = 40$ ksi, multiply table values by 1.25.

Deflection of concrete structures presents a number of special problems. For concrete with ordinary reinforcing (not pre-stressed), flexural action normally results in tension cracking of the concrete at points of maximum bending. Thus the presence of cracks in the bottom of a beam at midspan points and in the top over supports is to be expected. In general, the size (and visibil-ity) of these cracks will be proportionate to the amount of beam curvature produced by deflection. Crack size will also be greater for long spans and for deep beams. If visible cracking is consid-ered objectionable, more conservative depth/span ratios should be used, especially for spans over 30 ft and beam depths over 30 in.

Creep of concrete (see Section 2-11) results in additional de-flections over time. This is caused by the sustained loads—essen-tially the dead load of the construction. Deflection controls reflect concern for this as well as for the instantaneous deflection under live load, the latter being the major concern in structures of wood and steel.

In beams, deflections, especially creep deflections, may be re-duced by the use of some compressive reinforcing. Where deflec-

tions are of concern, or where depth/span ratios are pushed to their limits, it is advisable to use some compressive reinforcing, consisting of continuous top bars.

Three fairly recent developments in concrete technology have resulted in increased concern for deflections. With improvements in production processes and controls, it has become possible to use higher values of concrete strength as measured by the value of f'_c. While the resulting strength (stress limit) increase is generally beneficial, the use of higher stresses in design produces higher strains, resulting in increased deformations. In addition, it may be observed from the Code formula for determination of the modulus of elasticity (Section 2-10) that the value for E increases only as the square root of f'_c. Thus, if f'_c is increased from 3000 to 4000 psi—a stress limit increase of 33%—the increase in E is only about 15%. This latter effect means that deflections increase more rapidly than strength gains with an improvement in f'_c.

The second development has been the trend toward use of higher grades of reinforcing steel with greater values of f_y. As with increases of f'_c, this results in the utilization of higher stresses and the accompanying higher strains in the steel, contributing to additional deformation.

Finally, there has been a trend to greater use of synthetic aggregates (in place of ordinary gravel) that usually effect a reduction in the weight of the finished concrete. It is possible in most cases to effect a weight reduction of up to 30% while still maintaining a strength (f'_c) that is equivalent to that of ordinary weight concrete. Again, with reference to the formula for E (Section 2-10), it may be noted that reduction in weight results in loss of modulus of elasticity and thus in increased deformations. In this situation there is some compensation in the form of less dead load, which is particularly significant to creep deflections.

When, for whatever reasons, deflections are deemed to be critical, computations of actual values of deflection may be necessary. Section 9.5 of the ACI Code provides directions for such computations; they are quite complex in most cases, and beyond the scope of this work. In actual design work, however, they are required very infrequently.

9-6. Design Procedure for Rectangular Beams

Let us consider a design procedure for a rectangular section with tension reinforcing only. We assume the following as the design situation regarding the problem data.

Given. Maximum bending moment, maximum shear force, beam span length and end conditions, f'_c, f_y, required concrete cover for reinforcing, and size of stirrups (if any).

Required: The beam dimensions (b and h), the flexural reinforcing, and the shear reinforcing.

Working Stress Method. The following procedure can be used with the working stress method.

1. Find bd^2 required, using the relationship $M = Rbd^2$ (R from Table 6-2; formula from Section 6-4).
2. Find bd required, using the relationship $V = vbd$ with a value of $2v_c$ for v. (See Section 7-5.) A maximum shear stress of $2v_c$ will usually result in moderate quantities of shear reinforcing.
3. Find minimum h for deflection (See Tables 9-2 or 9-3).
4. Pick trial values for b and h.
5. Find A_s required.
   a. If actual bd^2 is approximately equal to the required M/R, use the value of p for a balanced section (from Table 6-2) and compute: $A_s = pbd$.
   b. If actual bd^2 exceeds M/R, guess at a true value for j. (Something larger than the balanced value in Table 6-2.) Then compute: $A_s = M/(f_s \times jd)$. Use this value of A_s to find the actual p and check j using Fig. 6-4. Verify that the actual percentage of reinforcing is not less than the required minimum of $200/f_y$.
6. Pick a set of bars. (See inside of back cover for bar properties.) Check the adequacy of the assumed b dimension with that given in Table 9-1.
7. Check for any problems involving development lengths. (See Sections 8-6 and 8-7.)
8. Design the shear reinforcing. (See Section 7-5.)

The following example illustrates the procedure.

Example 1. A simple beam has a 20-ft [6.10-m] span and carries a uniformly distributed load with a total value of 80 k [358.4 kN], resulting in a maximum bending moment of 200 k-ft [271.2 kN-m] and a maximum end shear of 40 k [179.2 kN]. A concrete beam is to be used with f'_c = 3 ksi [20.7 MPa] and steel with f_y = 40 ksi [276 MPa], cover of 1.5 in. [38 mm], and No. 3 U-stirrups. Find the dimensions for a rectangular section with tension reinforcing only and pick the bars for the flexural reinforcing.
Solution: Using Table 6-2, we find: j = 0.872, p = 0.0129, R = 0.226 in k-in. [1554 in kN-m]. Then, referring to the steps in the previously described procedure:

1. Required bd^2 = 200 × 12/0.226 = 10,619 in.^{3}

$$\left[\frac{271.2}{1554} = 0.1745 \text{ m}^3\right]$$

2. Allowable v_c for f'_c = 1.1 $\sqrt{f'_c}$ = 1.1 $\sqrt{3000}$ = 60 psi [414 kPa].

$$\text{required } bd = V/2v_c = 40,000/120 = 333 \text{ in.}^2$$

$$[179.2/828 = 0.217 \text{ m}^2]$$

3. For the simple span, Table 9-2 requires h = L/16. Thus

$$\text{minimum } h = 20 \times \frac{12}{16} = 15 \text{ in. [380 mm]}$$

We next proceed to consider combinations of b and d that will produce an approximately balanced section.

4. If b = 10 in., required d = $\sqrt{\dfrac{10,619}{10}}$ = 32.6 in.

$$\text{or } d = \frac{333}{10} = 33.3 \text{ in.}$$

This section is quite narrow and deep, so a larger value of b is indicated. Try

$$b = 14 \text{ in.}, \ d = \sqrt{\frac{10,619}{14}} = 27.5 \text{ in.}$$

$$\text{or} \quad d = \frac{333}{14} = 23.8 \text{ in.}$$

$$\left[b = 0.355 \text{ m}, \ d = \sqrt{\frac{0.1745}{0.355}} = 0.701 \text{ m}, \right.$$

$$\left. d = \frac{0.217}{0.355} = 0.611 \text{ m} \right]$$

Try

$$b = 14 \text{ in. } [0.355 \text{ m}], \ h = 30 \text{ in. } [0.760 \text{ m}]$$
$$d \approx h - 2.5 \text{ in.} = 27.5 \text{ in.}$$

$$[d = h - 0.063 \text{ m} = 697 \text{ m}]$$

5. Since the chosen section is approximately balanced, we may find the steel area as

$$A_s = pbd = 0.0129(14 \times 27.5) = 4.97 \text{ in.}^2$$

$$[0.0129(355 \times 697) = 3192 \text{ mm}^2]$$

6. Possible choices are

No.	Bar size	Actual A_s	Minimum width (see Table 9-1)
4	10	5.08 in.²	13 12 in.
5	9	5.00	14 in.

After the computation of the required dimensions for a balanced section, there may be some reason for choosing a larger section. If so, the procedure is as follows. If $b = 14$ in. [0.355 m] and $h = 36$ in. [0.915 m], then $d = 36 - 2.5 = 33.5$ in. [0.915 −

$0.063 = 0.852$ m]. Assume a larger value for j, say 0.90, and find A_s as follows:

$$A_s = \frac{M}{f_s jd} = \frac{200 \times 12}{20 \times 0.90 \times 33.5} = 3.98 \text{ in.}^2$$

$$\left[\frac{271.2 \times 10^3}{138 \times 0.90 \times 0.852} = 2563 \text{ mm}^2 \right]$$

With this value of A_s, the actual value of p is

$$p = \frac{A_s}{bd} = \frac{3.98}{14 \times 33.5} = 0.0085$$

The minimum value for p, as required by Section 10.5.1 of the ACI Code is

$$\text{minimum } p = \frac{200}{f_y} = \frac{200}{40,000} = 0.005$$

From Fig. 6-4 we find an approximate value for k to be

$$k = 0.320$$

Thus

$$j = 1 - \frac{k}{3} = 1 - \frac{0.320}{3} = 0.893$$

and the assumed value is reasonable.

Bar choice: 4 No. 9, actual $A_s = 4.00$ in.^{2} [2580 mm^2]

Strength Design Method.   The following is a procedure to be used with the strength design method.

1. Convert service loads to design loads. (See Section 4-9.) Thus, for gravity loads only,

$$M_u = 1.4M_d + 1.7M_l$$
$$V_u = 1.4V_d + 1.7V_l$$

2. Compute required design strengths using the capacity reduction factors.

$$M_t = \frac{M_u}{0.90}$$

$$V_t = \frac{V_u}{0.85}$$

3. Find p, a/d, and R from Table 6-3 for the minimum-size section with the maximum permitted reinforcing. Pick a trial b and find the required d and A_s, using

$$bd^2 - \frac{M_t}{R}, \quad A_s = pbd$$

4. Compute the minimum $p = 200/f_y$. Using this p, compute a/d and R from

$$\frac{a}{d} = p \frac{f_y}{0.85f'_c}$$

$$R = 0.85f'_c \left[\frac{a}{d} - 2 \left(\frac{a}{d} \right)^2 \right]$$

As for Step 3, pick b and find d and A_s.

If a section with a size between these limits is desired, proceed as follows:

5. Select the desired dimensions b and h (overall height) and estimate d based on the stirrup size and the required cover. (With a No. 3 U-stirrup and 1.5-in. cover, use a trial d of $h - 2.5$ in.) Then:
   a. Estimate a value for a/d based on the limits found in Steps 3 and 4.
   b. Find the required A_s from

$$A_s = \frac{M_t}{f_y(d - a/2)}$$

   c. Compute the value of p with this A_s and find the corresponding a/d as in Step 4.

   d. If a/d from Step 5c is significantly different from the estimate made in Step 5a, make a new estimate and repeat the process until a reasonable agreement is found between the estimated and computed values for a/d.

6. Pick the reinforcing and verify the beam width for spacing of the bars. (See Table 9-1.)

7. Design for shear. If a minimum-sized section is to be used (with maximum permitted p), it is best to check the feasibility of the section for shear before completing the design for flexure.

8. Investigate any problems involving development lengths.

The following example illustrates the procedure.

Example 2. Repeat the design in Example 1 using strength design. Assume the service load moment to be one-half dead load and one-half live load.
Solution:

1. The design moment is found as

$$M_u = 1.4M_d + 1.7M_l = 1.4(100) + 1.7(100)$$
$$= 310 \text{ k-ft } [430 \text{ kN-m}]$$

2. The required design strength is found as

$$M_t = \frac{M_u}{\phi} = \frac{M_u}{0.90} = \frac{310}{0.90}$$
$$= 345 \text{ k-ft } [478 \text{ kN-m}]$$

3. From Table 6-3, we find values of $p = 0.0278$, $a/d = 0.438$, and $R = 0.868$ [5985 in kN-m]. Try $b = 14$ in.; then required d is

$$d = \sqrt{\frac{M}{Rb}} = \sqrt{\frac{345 \times 12}{0.868 \times 14}} = 18.5 \text{ in.}$$

$$\left[b = 0.356 \text{ m}, d = \sqrt{\frac{478}{5985 \times 0.356}} = 0.474 \text{ m} \right]$$

If a section with these precise dimensions is chosen (b = 14 in., d = 18.5 in.), the value for p may be taken from the table; then

$$A_s = pbd = 0.0278(14 \times 18.5) = 7.20 \text{ in.}^2$$

$$[0.0278(356 \times 474) = 4691 \text{ mm}^2]$$

4. For the other limit, the minimum value for p is $200/f_y$ = $200/40,000 = 0.005$; and

$$\frac{a}{d} = p \times \frac{f_y}{0.85f'_c} = 0.005 \frac{40}{0.85(3)} = 0.0784$$

$$R = 0.85f'_c \left\{ \left(\frac{a}{d}\right) - \frac{1}{2} \left(\frac{a}{d}\right)^2 \right\}$$

$$= 0.85f'_c \left\{ (0.0784) - \frac{(0.0784)^2}{2} \right\}$$

$$= 0.192 \text{ (in k-in. units)}$$

$$\left[R = 0.85(20.7 \times 10^3)\left\{ (0.0784) - \frac{1}{2}(0.0784)^2 \right\} \right.$$
$$\left. = 1325 \text{ (in kN-m units)} \right]$$

Then,

$$bd^2 = \frac{M_t}{R}$$

and, if b = 14 in. [0.356 m]

$$d = \sqrt{\frac{M_t}{bR}} = \sqrt{\frac{345 \times 12}{14 \times 0.192}} = 39.2 \text{ in.}$$

$$\left[d = \sqrt{\frac{478}{0.356 \times 1325}} = 1.007 \text{ m} \right]$$

And, if these dimensions are used,

$$A_s = 0.005bd = 0.005(14 \times 39.2) = 2.74 \text{ in.}^2$$

$$[A_s = 0.005(356 \times 1007) = 1792 \text{ mm}^2]$$

5. If a section somewhere between these limits is desired, we first summarize what we know about the limiting conditions. Thus,

   when $b = 14$ in. and $d = 18.5$ in.,

   $$A_s = 7.20 \text{ in}^2 \text{ and } \frac{a}{d} = 0.438,$$

   when $b = 14$ in. and $d = 39.3$ in.,

   $$A_s = 2.74 \text{ in}^2 \text{ and } \frac{a}{d} = 0.0784$$

   Let us now try a section with true dimensions of $b = 14$ in. and $h = 30$ in. [0.356 m and 0.762 m]. Then:

   a. Assuming a value for d of $h - 2.5$ in., or 27.5 in., we first guess at an a/d value of 0.25. Thus:

   $a = 0.25d = 0.25(27.5) = 6.875$, say 7 in.

   $[d = 0.762 - 0.064 = 0.698$ m;

   $$a = 0.1745, \text{ say } 0.175 \text{ m}]$$

   b. $$A_s = \frac{M}{f_y(d - a/2)} = \frac{345 \times 12}{40(27.5 - 3.5)} = 4.31 \text{ in.}^2$$

   $$\left[A_s = \frac{478 \times 10^3}{276(0.698 - 0.175/2)} = 2937 \text{ mm}^2\right]$$

   c. $$p = \frac{A_s}{bd} = \frac{4.31}{14 \times 27.5} = 0.0112$$

   And, with this value for p,

   $$\frac{a}{d} = p\frac{f_y}{0.85f'_c} = (0.0112)\frac{40}{0.85(3)} = 0.176$$

   $a = 0.176d = 4.83$ in. [0.123 m]

   d. Since the computed value for a is considerably off the mark of the assumed value of 7 in., we try again.

Assume $a = 4.5$ in. [0.114 m], then

$$A_s = \frac{M}{f_y(d - a/2)} = \frac{345 \times 12}{40(27.5 - 2.25)}$$

$$= 4.10 \text{ in.}^2 \text{ [2645 mm}^2]$$

$$p = \frac{4.10}{14 \times 27.5} = 0.01065$$

$$\frac{a}{d} = p \frac{f_y}{0.85f'_c} = (0.01065) \frac{40}{0.85(3)} = 0.167$$

$$a = 0.167(27.5) = 4.59 \text{ in. } [0.117 \text{ m}]$$

It should be noted that the real concern in these computations is for a reasonable accuracy in the value for A_s. It may be observed that minor changes in the value of a result in even smaller changes in the value of A_s. Thus, if we are off as much as 10% in the guess of a, the effect on A_s is usually negligible.

Problem 9-6-A.   A simple beam has a 24-ft [7.32-m] span and carries a uniformly distributed load with a total value of 100 k, [445 kN], resulting in a maximum bending moment of 300 k-ft, [407 kN-m], and a maximum end shear of 50 k, [223 kN]. A concrete beam is to be used with $f'_c = 3$ ksi, [20.7 MPa], and steel with $f_y = 40$ ksi, [276 MPa], cover of 1.5 in. [38 mm], and No. 3 U-stirrups. Find the dimensions for a rectangular section by the working stress method, striving for an approximately balanced section. Pick the reinforcing bars for the section.

Problem 9-6-B.   Using the data from Problem 9-6-A, pick a section with dimensions slightly greater than those for a balanced section and determine the required reinforcing.

Problem 9-6-C.   Using the data from Problem 9-6-A, design the section using strength design methods.

9-7.   Design Procedure for T-Beams

When a floor slab and its supporting beams are poured at the same time, the result is a monolithic construction in which a portion of the slab on each side of the beam serves as the flange of a T-beam. The part of the section that projects below the slab is called the web or stem of the T-beam. This type of beam is shown in Fig. 9-3a. For positive moment, the flange is in compression and there

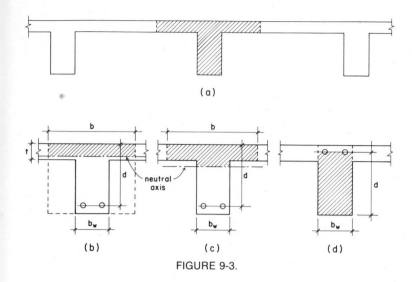

FIGURE 9-3.

is ample concrete to resist compressive stresses, as shown in Fig. 9-3*b* or 9-3*c*. However, in a continuous beam, there are negative bending moments over the supports, and the flange here is in the tension stress zone with compression in the web.

It is important to remember that only the area formed by the width of the web b_w and the effective depth d is to be considered in computing resistance to shear and to bending moment over the supports. This is the hatched area shown in Fig. 9-3*d*.

The effective flange width to be used in the design of symmetrical T-beams is limited to one-fourth the span length of the beam. In addition, the overhanging width of the flange on either side of the web is limited to eight times the thickness of the slab or one-half the clear distance to the next beam.

In monolithic construction with beams and one-way solid slabs, the effective flange area of the T-beams is usually quite capable of resisting the compressive stresses caused by positive bending moments. With a large flange area, as shown in Fig. 9-4, the neutral axis of the section usually occurs quite high in the beam web, resulting in only minor compressive stresses in the web. If the compression developed in the web is ignored, the net

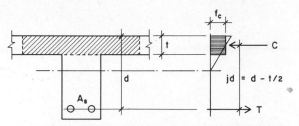

FIGURE 9-4. Basis for simplified analysis of a T-beam

compression force may be considered to be located at the centroid of the trapezoidal stress zone that represents the stress distribution in the flange. On this basis, the compression force is located at something less than $t/2$ from the top of the beam.

An approximate analysis of the T-section by the working stress method that avoids the need to find the location of the neutral axis and the centroid of the trapezoidal stress zone, consists of the following steps.

1. Ignore compression in the web and assume a constant value for compressive stress in the flange (see Fig. 9-4). Thus

$$jd = d - \frac{t}{2}$$

2. Find the required steel area as

$$A_s = \frac{M}{f_s jd} = \frac{M}{f_s(d - t/2)}$$

3. Check the compressive stress in the concrete as

$$f_c = \frac{C}{b_f \times t}, \qquad \text{where } C = \frac{M}{jd} = \frac{M}{d - t/2}$$

The actual value of maximum compressive stress will be slightly higher, but will not be critical if this computed value is significantly less than the limit of $0.45f'_c$.

The following example illustrates the use of this procedure. It assumes a typical design situation in which the dimensions of the

section (b_f, b_w, d, and t) are all predetermined by other design considerations and the design of the T-section is reduced to the requirement to determine the area of tension reinforcing.

Example 1. A T-section is to be used for a beam to resist positive moment. The following data is given: beam span = 18 ft [5.49 m], beams are 9 ft [2.74 m] center to center, slab thickness is 4 in. [0.102 m], beam stem dimensions are b_w = 15 in. [0.381 m] and d = 22 in. [0.559 m], f'_c = 4 ksi [27.6 MPa], f_y = 60 ksi [414 MPa], f_s = 24 ksi [165 MPa]. Find the required area and pick reinforcing bars for a dead load moment of 100 k-ft [136 kN-m] plus a live load moment of 100 k-ft [136 kN-m].

Solution: Using working stress design with the approximate method described previously:

1. Determine the effective flange width (necessary only for a check on the concrete stress). Maximum value for the flange width is

$$b_f = \frac{\text{span}}{4} = \frac{18 \times 12}{4} = 54 \text{ in. [1.37 m]}$$

or b_f = center-to-center beam spacing

$$= 9 \times 12 = 108 \text{ in. [2.74 m]}$$

or b_f = beam stem width plus 16 times the slab thickness

$$= 15 + 16(4) = 79 \text{ in. [2.01 m]}$$

The limiting value is therefore 54 in. [1.37 m].

2. Find the required steel area

$$A_s = \frac{M}{f_s(d - t/2)} = \frac{200 \times 12}{24(22 - 4/2)} = 5.00 \text{ in.}^2 \text{ [3364 mm}^2\text{]}$$

3. Pick bars using the table of properties inside the back cover and check the adequacy of the stem width using Table 9-1.
   From the properties table: Choose five No. 9 bars, actual A_s = 5.00 in.^{2}
   From Table 9-1: required width for five No. 9 bars is 14 in.; less than the 15 in. provided.

4. Check the concrete stress.

$$C = \frac{M}{jd} = \frac{200 \times 12}{20} = 120 \text{ k } [535 \text{ kN}]$$

$$f_c = \frac{C}{b_f t} = \frac{120}{54 \times 4} = 0.556 \text{ ksi } [3.83 \text{ MPa}]$$

limiting stress $= 0.45 f'_c = 0.45(4) = 1.8 \text{ ksi } [12.4 \text{ MPa}]$

Thus compressive stress in the flange is clearly not critical.

In a real design situation, of course, consideration would have to be given to problems of shear and possibly to problems of development lengths for the bars.

When using strength design methods for T-sections, we recommend a procedure similar to that described for the working stress method. This method and procedure assumes that the flange area of the T is so large that the concrete stress never gets up to its ultimate limit before the yield stress develops in the reinforcing. The following example illustrates the procedure.

Example 2.   Perform the design for the beam described in Example 1 using strength design methods.
Solution:

1. As in Example 1, effective $b_f = 54$ in. [1.37 m].
2. Design moment is found as

$$M_u = 1.4M_d + 1.7M_l$$

$$= 1.4(100) + 1.7(100)$$

$$= 310 \text{ k-ft } [420 \text{ kN-m}]$$

3. Required design strength is found as

$$M_t = \frac{M_u}{0.90} = \frac{310}{0.90} = 345 \text{ k-ft } [467 \text{ kN-m}]$$

4. Assuming the location of the net compression force to be at the center of the flange area, as described for the working stress method, the steel area is found as

$$A_s = \frac{M}{f_y(d - t/2)} = \frac{345 \times 12}{60 \times 20} = 3.45 \text{ in.}^2 \text{ [2221 mm}^2]$$

5. A possible choice for the reinforcing is two No. 8 bars plus two No. 9 bars, providing an actual area of 3.58 in.^{2} [2310 mm^2]. Table 9-1 shows that a minimum width for four No. 9 bars is only 12 in., so the stem width is more than adequate for bar spacing.
6. Assuming the steel bars to be stressed to the yield point, the average stress in the flange would be as follows.

$$C = T = f_y \times A_s = 60 \times 3.58 = 215 \text{ k [956 kN]}$$

$$f_c = \frac{C}{b_f t} = \frac{215}{54 \times 4} = 0.995 \text{ ksi [6.86 MPa]}$$

which is considerably lower than f'_c.

The examples in this section illustrate procedures that are reasonably adequate for beams that occur in ordinary beam and slab construction. When special T-sections occur with thin flanges (t less than $d/8$ or so) or narrow effective flange widths (b_f less than three times b_w or so), these methods may not be valid. In such cases more accurate investigation should be performed, using the requirements of the ACI Code.

Problem 9-7-A. A T-section is to be used for a beam to resist positive moment. The following data is given: beam span = 24 ft [7.32 m], beams are 10 ft [3.05 m] center to center, slab thickness is 5 in. [13 mm], beam stem dimensions are b_w = 18 in. [46 mm] and d = 30 in. [76 mm], f'_c = 4 ksi [27.6 MPa], f_y = 60 ksi [414 MPa], f_s = 24 ksi [165 MPa]. Find the required steel area and pick bars for a dead load moment of 160 k-ft [217 kN-m] plus a live load moment of 120 k-ft [165 kN-m]. Use the working stress method.

Problem 9-7-B. Using the data from Problem 9-7-A, design the beam using strength design methods.

9-8. One-Way Solid Slabs

One of the most commonly used concrete floor systems consists of a solid slab that is continuous over parallel supports. The supports may consist of bearing walls of masonry or concrete but most often consist of sets of evenly spaced concrete beams. The beams are usually supported by girders, which in turn are supported by columns. In this type of slab the principal reinforcement runs in one direction, parallel to the slab span and perpen-

dicular to the supports. For this reason it is called a *one-way solid slab*. The number and spacing of supporting beams depends on their span, the column spacing, and the magnitude of the loads. Most often the beams are spaced uniformly and frame into the girders at the center, third, or quarter points. The formwork for this type of floor is readily constructed, and the one-way slab is most economical for medium and heavy floor loads for relatively short spans, 6 to 12 ft. For long spans the slab thickness must be increased, resulting in considerable dead weight of the construction, which increases the cost of the slab and its reinforcing as well as the cost of supporting beams, girders, columns, and foundations.

To design a one-way slab, we consider a strip 12 in. wide (see Fig. 9-5). This strip is designed as a beam whose width is 12 in. and on which is a uniformly distributed load. As with any rectangular beam, the effective depth and tensile reinforcement are computed as explained in Section 6-4. A minimum slab thickness is often determined on the basis of the fire rating requirements of the applicable building code. A minimum thickness is also required to prevent excessive deflection. Based on deflection limitations, slab thicknesses should not be less than those given in Table 9-2.

9-9. Shrinkage and Temperature Reinforcement for Slabs

While flexural reinforcement is required in only one direction in the one-way slab, reinforcement at right angles to the flexural reinforcement is also provided for stresses due to shrinkage of the concrete and temperature fluctuations. The amount of this reinforcement is specified as a percentage p of the gross cross-sectional area of the concrete, as follows:

for slabs reinforced with Grade 40 or Grade 50 deformed bars,

$$p = \frac{A_s}{b \times t} = 0.0020$$

and for slabs reinforced with Grade 60 deformed bars,

$$p = 0.0018$$

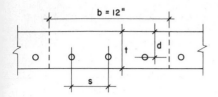

FIGURE 9-5.   Reference for slab design.

Center-to-center bar spacing must not be greater than five times the slab thickness or 18 in.

9-10.   Design of a One-Way Solid Slab

As discussed in Section 9-8, the one-way slab is designed by assuming the slab to consist of a series of 12-in. [305-mm] wide segments. The bending moment for this 12-in.-wide rectangular element is determined, and the required effective depth and area of tensile reinforcement A_s is computed using the procedures for a rectangular beam as discussed in Section 6-4. The A_s thus determined is the average amount of steel per 1 ft width of slab that is required. The maximum spacing (s in Fig. 9-5) for this reinforcement is three times the slab thickness or a maximum of 18 in. [457 mm]. The size and spacing of bars may be selected by use of Table 9-4.

It is not practicable to use shear reinforcement in one-way slabs, and consequently the maximum unit shear stress must be kept within the limit for the concrete alone. The usual procedure is to check the shear stress with the effective depth determined for bending before proceeding to find A_s. Except for very short span slabs with excessively heavy loadings, shear stress is seldom critical.

The following example illustrates the design procedure for a one-way solid slab.

Example.   A one-way solid concrete slab is to be used for a simple span of 14 ft [4.27 m]. In addition to its own weight, the slab carries a superimposed dead load of 30 psf [1.44 kN/m²] and a live load of 100 psf [4.79 kN/m²]. Using f'_c = 3 ksi [20.7 MPa], f_y = 40 ksi [276 MPa], and f_s = 20 ksi [138 MPa], design the slab for minimum overall thickness.

TABLE 9-4. Areas of Bars in Reinforced Concrete Slabs per Foot of Width

Spacing (in.)	Areas of bars (in.²)									
	No. 2	No. 3	No. 4	No. 5	No. 6	No. 7	No. 8	No. 9	No. 10	No. 11
3	0.20	0.44	0.79	1.23	1.77	2.41	3.14	4.00		
3½	0.17	0.38	0.67	1.05	1.51	2.06	2.69	3.43	4.36	
4	0.15	0.33	0.59	0.92	1.33	1.80	2.36	3.00	3.81	4.68
4½	0.13	0.29	0.52	0.82	1.18	1.60	2.09	2.67	3.39	4.16
5	0.12	0.26	0.47	0.74	1.06	1.44	1.88	2.40	3.05	3.74
5½	0.11	0.24	0.43	0.67	0.96	1.31	1.71	2.18	2.77	3.40
6	0.10	0.22	0.39	0.61	0.88	1.20	1.57	2.00	2.54	3.12
6½	0.09	0.20	0.36	0.57	0.82	1.11	1.45	1.85	2.35	2.88
7	0.08	0.19	0.34	0.53	0.76	1.03	1.35	1.71	2.18	2.67
7½	0.08	0.18	0.31	0.49	0.71	0.96	1.26	1.60	2.03	2.50
8	0.07	0.17	0.29	0.46	0.66	0.90	1.18	1.50	1.91	2.34
8½	0.07	0.16	0.28	0.43	0.62	0.85	1.11	1.41	1.79	2.20
9	0.07	0.15	0.26	0.41	0.59	0.80	1.05	1.33	1.69	2.08
9½	0.06	0.14	0.25	0.39	0.56	0.76	0.99	1.26	1.60	1.97
10	0.06	0.13	0.24	0.37	0.53	0.72	0.94	1.20	1.52	1.87
11	0.05	0.12	0.21	0.33	0.48	0.66	0.86	1.09	1.39	1.70
12	0.05	0.11	0.20	0.31	0.44	0.60	0.79	1.00	1.27	1.56

Solution:
Working Stress Method

Using the general procedure for design of a beam with rectangular section (Section 6-5), we first determine the required slab thickness. Thus

For deflection, from Table 9-2,

$$\text{minimum } t = \frac{L}{25} = \frac{14 \times 12}{25} = 6.72 \text{ in. [171 mm]}$$

For flexure we first determine the maximum bending moment. The loading must include the weight of the slab, for which we use the thickness required for deflection as a first estimate. Assuming a 7-in. [178-mm] thick slab, then slab weight is $\frac{7}{12}$ (150 pcf) = 87.5 psf, say 88 psf and total load is 100 psf LL + 118 psf DL = 218 psf.

The maximum bending moment for a 12-in.-wide design strip of the slab thus becomes

$$M = \frac{w\,L^2}{8} = \frac{218(14)^2}{8} = 5341 \text{ ft-lb } [7.24 \text{ kN-m}]$$

For minimum slab thickness, we consider the use of a balanced section, for which Table 6-2 yields the following properties.

$$j = 0.872,\ p = 0.0129,\ R = 0.226$$

Then

$$bd^2 = \frac{M}{R} = \frac{5.341 \times 12}{0.226} = 284 \text{ in.}^3$$

And, since b is the 12-in. design strip width,

$$d = \sqrt{\frac{284}{12}} = \sqrt{23.7} = 4.86 \text{ in. } [123 \text{ mm}]$$

Assuming an average bar size of a No. 6 ($\frac{3}{4}$ in. nominal diameter) and cover of $\frac{3}{4}$ in., the minimum required slab thickness based on flexure becomes

$$t = 4.86 + \frac{0.75}{2} + 0.75 = 5.985 \text{ in.}$$

$$\left[123 + \frac{19}{2} + 19 = 152 \text{ mm} \right]$$

We thus observe that the deflection limitation controls in this situation, and the minimum overall thickness is the 6.72-in. dimension. If we continue to use the 7-in. overall thickness, the actual effective depth with a No. 6 bar will be

$$d = 7.0 - 1.125 = 5.875 \text{ in.}$$

Since this d is larger than that required for a balanced section, the value for j will be slightly larger than 0.872, as found from Table 6-2. Let us assume a value of 0.9 for j and determine the required area of reinforcement as

$$A_s = \frac{M}{f_s\,jd} = \frac{5.341 \times 12}{20(0.9)5.875} = 0.606 \text{ in.}^2$$

From Table 9-4, we find that the following bar combinations will satisfy this requirement:

Bar size	Spacing from center to center (in.)	Average A_s in a 12-in. width
No. 5	6	0.61
No. 6	8.5	0.62
No. 7	12	0.60
No. 8	15	0.63

The ACI Code requires a maximum spacing of three times the slab thickness (21 in. in this case). Minimum spacing is largely a matter of the designer's judgment. Many designers consider a minimum practical spacing to be one approximately equal to the slab thickness. Within these limits, any of the bar size and spacing combinations listed are adequate.

As described in Section 9-9, the ACI Code requires a minimum reinforcement for shrinkage and temperature effects to be placed in the direction perpendicular to the flexural reinforcement. With the Grade 40 bars in this example, the minimum percentage of this steel is 0.0020, and the steel area required for a 12-in. strip thus becomes

$$A_s = p(b \times t) = 0.0020(12 \times 7) = 0.168 \text{ in.}^2$$

From Table 9-4, we find that this requirement can be satisfied with No. 3 bars at 8-in. centers or No. 4 bars at 14-in. centers. Both of these spacings are well below the maximum of five times the slab thickness.

Although simply supported single slabs are sometimes encountered, the majority of slabs used in building construction are continuous through multiple spans. An example of the design of such a slab is given in the next chapter.

Strength Design Method

Strength design procedures for the slab are essentially the same as for the rectangular beam, as described in Section 6-7. In most cases, slab sections will be reinforced with steel areas well

below those for a balanced section, so the procedure for a so-called underreinforced section should be used. If the procedure illustrated in Section 6-7 is used for this example, it will be found that the required steel area is approximately 15% less than that required from the working stress method computations.

Problem 9-10-A.   A one-way solid concrete slab is to be used for a simple span of 16 ft [4.88 m]. In addition to its own weight, the slab carries a superimposed dead load of 60 psf [2.87 kN/m²] and a live load of 75 psf [3.59 kN/m²]. Using the working stress method with f'_c = 3 ksi [20.7 MPa], f_y = 40 ksi [276 MPa], and f_s = 20 ksi [138 MPa], design the slab for minimum overall thickness.

Problem 9-10-B.   Using the data from Problem 9-10-A, design the slab using strength design methods.

9-11. Beams with Compressive Reinforcement

There are many situations in which steel reinforcing is used on both sides of the neutral axis in a beam. When this occurs, the steel on one side of the axis will be in tension and that on the other side in compression. Such a beam is referred to as a doubly reinforced beam or simply as a beam with compressive reinforcing (it being naturally assumed that there is also tensile reinforcing). Various situations involving such reinforcing have been discussed in the preceding sections. In summary, the most common occasions for such reinforcing include:

1. The desired resisting moment for the beam exceeds that for which the concrete alone is capable of developing the necessary compressive force.

2. Other functions of the section require the use of reinforcing on both sides of the beam. These include the need for bars to support U-stirrups and situations when torsion is a major concern.

3. It is desired to reduce deflections by increasing the stiffness of the compressive side of the beam. This is most significant for reduction of long-term creep deflections.

4. The combination of loading conditions on the structure result in reversal moments on the section. That is, the section must sometimes resist positive moment, and other times resist negative moment.

5. Anchorage requirements (for development of reinforcing) require that the bottom bars in a continuous beam be extended a significant distance into the supports.

The precise investigation and accurate design of doubly reinforced sections, whether performed by the working stress or by strength design methods, is quite complex and is beyond the scope of work in this book. The following discussion presents an approximation method that is adequate for preliminary design of a doubly reinforced section. For real design situations, this method may be used to establish a first trial design, which may then be more precisely investigated using more rigorous methods.

For the beam with double reinforcing, as shown in Fig. 9-6, we consider the total resisting moment for the section to be the sum of the following two component moments.

M_1 (Fig. 9-6b) is comprised of a section with tension reinforcing only (A_{s1}). This section is subject to the usual procedures for design and investigation, as discussed in Sections 6-5 and 9-6.

M_2 (Fig. 9-6c) is comprised of two opposed steel areas (A_{s2} and A_s') that function in simple moment couple action, similar to the flanges of a steel beam or the top and bottom chords of a truss.

Ordinarily, we expect that $A_{s2} = A_s'$, since the same grade of steel is usually used for both. However, there are two special considerations that must be made. The first involves the fact that

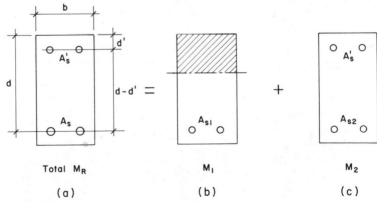

FIGURE 9-6. Basis for simplified analysis of a doubly-reinforced section.

A_{s2} is in tension, while A_s' is in compression. A_s' must therefore be dealt with in a manner similar to that for column reinforcing, as discussed in Chapter 11. This requires, among other things, that the compressive reinforcing be braced against buckling, using ties similar to those in a tied column.

The second consideration involves the distribution of stress and strain on the section. Referring to Fig. 9-7, it may be observed that, under normal circumstances (kd less than $0.5d$), A_s' will be closer to the neutral axis than A_{s2}. Thus the stress in A_s' will be lower than that in A_{s2} if pure elastic conditions are assumed. However, it is common practice to assume steel to be doubly stiff when sharing stress with concrete in compression, due to shrinkage and creep effects. Thus, in translating from linear strain conditions to stress distribution, we use the relation $f_s'/2n$ (where $n = E_s/E_c$, as discussed in Section 2-10). Utilization of this relationship is illustrated in the following examples.

Example 1. A concrete section with $b = 18$ in. [0.457 m] and $d = 21.5$ in. [0.546 m] is required to resist service load moments as follows: dead load moment $= 150$ k-ft [203.4 kN-m], live load moment $= 150$ k-ft [203.4 kN-m]. Using working stess methods, find the required reinforcing. Use $f_c' = 4$ ksi [27.6 MPa] and $f_y = 60$ ksi [414 MPa].

Solution: For the Grade 60 reinforcing, we use an allowable stress of $f_s = 24$ ksi [165 MPa]. Then, using Table 6-2, find:

$$n = 8, \quad k = 0.375, \quad j = 0.875, \quad p = 0.0141$$

$$R = 0.295 \text{ in k-in. units } [2028 \text{ in kN-m units}]$$

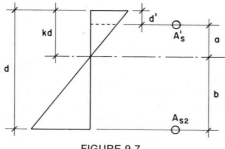

FIGURE 9-7.

Using the R value for the balanced section, the maximum resisting moment of the section is

$$M = Rbd^2 = \frac{0.295}{12} \times (18)(21.5)^2$$

$$= 205 \text{ k-ft } [278 \text{ kN-m}]$$

This is M_1, as shown in Fig. 9-6. Thus

$$M_2 = \text{total } M - M_1 = 300 - 205 = 95 \text{ k-ft}$$

$$[407 - 278 = 129 \text{ kN-m}]$$

For M_1 the required reinforcing (A_{s1} in Fig. 9-6) may be found as

$$A_{s1} = pbd = 0.0141 \times 18 \times 21.5 = 5.46 \text{ in.}^2 \text{ [3523 mm}^2]$$

And, assuming $f'_s = f_s$, we find A'_s and A_{s2} as follows.

$$M_2 = A'_s(d - d') = A_{s2}(d - d')$$

$$A'_s = A_{s2} = \frac{M_2}{f_s(d - d')} = \frac{95 \times 12}{24 \times 19} = 2.50 \text{ in.}^2 \text{ [1613 mm}^2].$$

The total tension reinforcing is thus

$$A_s = A_{s1} + A_{s2} = 5.46 + 2.50 = 7.96 \text{ in.}^2 \text{ [5136 mm}^2]$$

For the compressive reinforcing, we must find the proper limit for f'_s. To do this, we assume the neutral axis of the section to be that for the balanced section, producing the situation that is shown in Fig. 9-8. Based on this assumption, the limit for f'_s is found as follows.

$$\frac{f'_s}{2n} = \frac{5.56}{8.06} (0.45 \times 4) = 1.24 \text{ ksi}$$

$$f'_s = 2n \times 1.24 = 2 \times 8 \times 1.24 = 19.84 \text{ ksi } [137 \text{ MPa}]$$

Since this is less than the limit of 24 ksi, we must use it to find A'_s; thus

$$A' = \frac{M_2}{f'_s(d - d')} = \frac{95 \times 12}{19.84 \times 19} = 3.02 \text{ in.}^2 \text{ [1948 mm}^2]$$

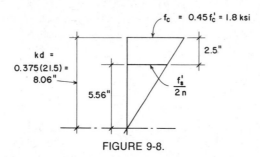

FIGURE 9-8.

In practice, compressive reinforcing is often used even when the section is theoretically capable of developing the necessary resisting moment with tension reinforcing only. The following two examples illustrate procedures that are applicable in this situation.

Example 2. Design the beam in Example 1 using strength design methods.

Solution: We first find the design moment in the usual manner.

$$M_u = 1.4M_d + 1.7M_l = 1.4(150) + 1.7(150)$$
$$= 465 \text{ k-ft } [631 \text{ kN-m}]$$

$$M_t = \frac{M_u}{\phi} = \frac{465}{0.9} = 517 \text{ k-ft } [701 \text{ kN-m}]$$

As a point of reference, we next determine the maximum resisting moment for the section with tension reinforcing only. Thus, using Table 6-3, we find

$$R = 1.041, \ p = 0.0214, \ \frac{a}{d} = 0.3773$$

and

$$M = Rbd^2 = \frac{1.041}{12} (18)(21.5)^2 = 722 \text{ k-ft } [979 \text{ kN-m}]$$

This indicates that the section could actually function without compressive reinforcing. However, we will assume that there are compelling reasons for having some compressive reinforcing al-

though its *amount* (magnitude of A'_s) becomes somewhat arbitrary. As a rough guide, we suggest a trial design with A'_s approximately one-third of A_s. On the basis of the previous computation, we know that the value for A_s will be less than that required for the full maximum resisting moment. That is, A_s will be less than

$$A_s = pbd = (0.0214)(18)(21.5) = 8.28 \text{ in.}^2 \text{ [5342 mm}^2]$$

For a trial design, we choose compressive reinforcing consisting of two No. 9 bars, with $A'_s = 2.0 \text{ in.}^2$ [1290 mm²]. With this reinforcing we may now compute a value for M_2, but to do so we must first establish a value for f'_s, the usable stress in the compressive reinforcing. For an approximate design, we may use the relationship shown in Fig. 9-9, in which we visualize the limit for f'_s to be $2n$ times the maximum stress of $0.85 f'_c$ in the concrete. Thus

$$f'_s = 2n(0.85 f'_c) = 2 \times 8 \ (0.85 \times 4) = 54.4 \text{ ksi [375 MPa]}$$

Since this value is less than the limiting yield strength, we use it to find M_2, thus

$$M_2 = A'_s f'_s (d - d')$$
$$= 2.0 \times 54.4 \times 19 \times \tfrac{1}{12} = 172 \text{ k-ft [233 kN-m]}$$

A_{s2} will have a value different from A'_s, since the value for stress for A_{s2} will be the full yield stress of 60 ksi. Thus

$$M_2 = A_{s2}f_y\ (d - d') = 172 \text{ k-ft}$$

$$A_{s2} = \frac{172 \times 12}{60\ (19)} = 1.81 \text{ in.}^2 \text{ [1168 mm}^2]$$

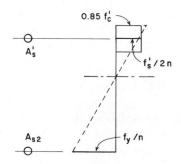

FIGURE 9-9.

With the value of M_2 established, we now find the required value for M_1. Thus

$$M_1 = M_t - M_2 = 517 - 172 = 345 \text{ k-ft}$$

$$[701 - 233 = 468 \text{ kN-m}]$$

To find the required value for A_{s1}, we use the usual procedure for a section with tension reinforcing only, as described in Section 9-6. Since the required value for M_1 is almost half of the maximum resisting moment (722 k-ft, as previously computed), we may assume that a/d will be considerably smaller than the table value of 0.377. For a first guess try

$$\frac{a}{d} = 0.20, \, a = 0.20(21.5) = 4.3 \text{ in.}$$

Rounding this off to 4 in., we find

$$A_{s1} = \frac{M_1}{f_y(d - a/2)} = \frac{345 \times 12}{60 \times 19.5} = 3.53 \text{ in.}^2 \, [2277 \text{ mm}^2]$$

With this area of steel, $p = 3.53/(18 \times 21.5) = 0.00912$ and

$$a = pd \frac{f_y}{0.85 f_c'} = (0.00912)(21.5) \frac{60}{0.85 \times 4}$$

$$= 3.46 \text{ in.}$$

For a second try, guess $a = 3.4$ in. Then

$$A_{s1} = \frac{345 \times 12}{60 \times 19.8} = 3.48 \text{ in.}^2$$

and

$$\text{total } A_s = A_{s1} + A_{s2} = 3.48 + 1.81 = 5.29 \text{ in.}^2$$

With these computations completed, we now make a choice of reinforcing for the section as follows.

compressive reinforcing
two No. 9 bars, $A_s' = 2.0$ in.^{2} [1290 mm^2]

tensile reinforcing
two No. 10 + two No. 11 bars, $A_s = 5.66$ in.^{2} [3652 mm^2]

The following example illustrates a procedure that may be used with the working stress method, when the required resisting moment is less than the balanced section limiting moment. It is generally similar to the procedure used with strength design in Example 2.

Example 3. Design a section by the working stress method for a moment of 180 k-ft [244 kN-m]. Use the section dimensions and data given in Example 1.

Solution: The first step is to investigate the section for its balanced stress limiting moment, as was done in Example 1. This will show that the required moment is less than the balanced moment limit, and that the section could function without compressive reinforcing. Again, we assume that compressive reinforcing is desired, so we assume an arbitrary amount for A_s' and proceed as in Example 2. We make a first guess for the total tension reinforcing as

$$A_s = \frac{M}{f_s(0.9d)} = \frac{180 \times 12}{24(0.9 \times 21.5)} = 4.65 \text{ in.}^2 \text{ [3000 mm}^2]$$

Try

$$A_s' = \tfrac{1}{3} A_s = \tfrac{1}{3}(4.65) = 1.55 \text{ in.}^2 \text{ [1000 mm}^2]$$

Choose two No. 8 bars,

$$\text{actual } A_s' = 1.58 \text{ in.}^2 \text{ [1019 mm}^2]$$

Thus

$$A_{s1} = A_s - A_s' = 4.65 - 1.58 = 3.07 \text{ in.}^2 \text{ [1981 mm}^2]$$

Using A_{s1} for a rectangular section with tension reinforcing only (See Section 6-4.)

$$p = \frac{3.07}{18 \times 21.5} = 0.0079$$

Then, from Fig. 6-4, we find $k = 0.30, j = 0.90$.

Using these values for the section, and the formula involving the concrete stress in compression from Section 6-4, we find

$$f_c = \frac{2M_1}{kjbd^2} = \frac{2 \times 120 \times 12}{0.3 \times 0.9 \times 18 \times (21.5)^2}$$

$$= 1.28 \text{ ksi } [8.83 \text{ MPa}]$$

With this value for the maximum concrete stress and the value of 0.30 for k, the distribution of compressive stress will be as shown in Fig. 9-10. Form this, we determine the limiting value for f_s' as follows.

$$\frac{f_s'}{2n} = \frac{3.95}{6.45}(1.28) = 0.784 \text{ ksi}$$

$$f_s' = 2n(0.784) = 2 \times 8 \times 0.784 = 12.5 \text{ ksi } [86.2 \text{ MPa}]$$

Since this is lower than f_s, we use it to find the limiting value for M_2. Thus

$$M_2 = A_s'f_s' (d - d')$$

$$= (1.58)(12.5)(19) \frac{1}{12} = 31 \text{ k-ft } [42 \text{ kN-m}]$$

To find A_{s2} we use this moment with the full value of $f_s = 24$ ksi. Thus

$$A_{s2} = \frac{M_2}{f_s(d - d')} = \frac{31 \times 12}{24 \times 19.0} = 0.82 \text{ in.}^2 [529 \text{ mm}^2]$$

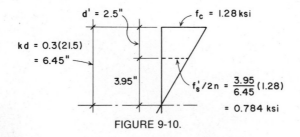

FIGURE 9-10.

To find A_{s1}, we determine that

$$M_1 = \text{total } M - M_2 = 180 - 31 = 149 \text{ k-ft } [202 \text{ kN-m}]$$

$$A_{s1} = \frac{M_1}{f_s jd} = \frac{149 \times 12}{24 \times 0.9 \times 21.5} = 3.85 \text{ in.}^2 \ [2484 \text{ MPa}]$$

Then the total tension reinforcing is found as

$$A_s = A_{s1} + A_{s2} = 3.85 + 0.82 = 4.67 \text{ in.}^2 \ [3013 \text{ mm}^2]$$

Problem 9-11-A. A concrete section with b = 16 in. [0.406 m] and d = 19.5 in. [0.495 m] is required to resist service load moments as follows: dead load moment = 120 k-ft [163 kN-m], live load moment = 110 k-ft [136 kN-m]. Using working stress methods, find the required reinforcing. Use f'_c = 4 kis [27.6 MPa] and Grade 60 bars with f_y = 60 ksi [414 PMa] and f_s = 24 ksi [165 MPa].

Problem 9-11-B. Using data from Problem 9-11-A, design a doubly reinforced section using strength methods. Use compressive reinforcing to develop approximately one-third of the required moment.

10

Reinforced Concrete Floor Systems
||||||||||||||||||||||||||||||||||||||||||||||||

10-1. Introduction

There are many different reinforced concrete floor systems, both cast in place and precast. The cast-in-place systems are generally of one of the following types:

1. One-way solid slab and beam.
2. Two-way solid slab and beam.
3. One-way concrete joist construction.
4. Two-way flat slab or flat plate without beams.
5. Two-way joist construction, called waffle construction.

Each system has its distinct advantages and limitations, depending on the spacing of supports, magnitude of loads, required fire rating, and cost of construction. The floor plan of the building and the purpose for which the building is to be used determine loading conditions and the layout of supports. Whenever possible, columns should be aligned in rows and spaced at regular intervals

TABLE 10-1. Weights of Building Construction

	lb/ft^2	kN/m^2
Roofs		
3-ply ready roofing (roll, composition)	1	0.05
3-ply felt and gravel	5.5	0.26
5-ply felt and gravel	6.5	0.31
Shingles		
Wood	2	0.10
Asphalt	2–3	0.10–0.15
Clay tile	9–12	0.43–0.58
Concrete tile	8–12	0.38–0.58
Slate, $\frac{1}{4}$ in.	10	0.48
Fiber glass	2–3	0.10–0.15
Aluminum	1	0.05
Steel	2	0.10
Insulation		
Fiber glass batts	0.5	0.025
Rigid foam plastic	1.5	0.075
Foamed concrete, mineral aggregate	2.5/in.	0.0047/mm
Wood rafters		
2 × 6 at 24 in.	1.0	0.05
2 × 8 at 24 in.	1.4	0.07
2 × 10 at 24 in.	1.7	0.08
2 × 12 at 24 in.	2.1	0.10
Steel deck, painted		
22 gage	1.6	0.08
20 gage	2.0	0.10
18 gage	2.6	0.13
Skylight		
Glass with steel frame	6–10	0.29–0.48
Plastic with aluminum frame	3–6	0.15–0.29
Plywood or softwood board sheathing	3.0/in.	0.0057/mm
Ceilings		
Suspended steel channels	1	0.05
Lath		
Steel mesh	0.5	0.025
Gypsum board, $\frac{1}{2}$ in.	2	0.10
Fiber tile	1	0.05
Dry wall, gypsum board, $\frac{1}{2}$ in.	2.5	0.12
Plaster		
Gypsum, acoustic	5	0.24
Cement	8.5	0.41
Suspended lighting and air distribution systems, average	3	0.15

TABLE 10-1. (*Continued*)

	lb/ft²	kN/m²
Floors		
Hardwood, ½ in.	2.5	0.12
Vinyl tile, ⅛ in.	1.5	0.07
Asphalt mastic	12/in.	0.023/mm
Ceramic tile		
¾ in.	10	0.48
Thin set	5	0.24
Fiberboard underlay, ⅝ in.	3	0.15
Carpet and pad, average	3	0.15
Timber deck	2.5/in.	0.0047/mm
Steel deck, stone concrete fill, average	35–40	1.68–1.92
Concrete deck, stone aggregate	12.5/in.	0.024/mm
Wood joists		
2 × 8 at 16 in.	2.1	0.10
2 × 10 at 16 in.	2.6	0.13
2 × 12 at 16 in.	3.2	0.16
Lightweight concrete fill	0.8/in.	0.015/mm
Walls		
2 × 4 studs at 16 in., average	2	0.10
Steel studs at 16 in., average	4	0.20
Lath, plaster; see Ceilings		
Gypsum dry wall, ⅝ in. single	2.5	0.12
Stucco, ⅞ in., on wire and paper or felt	10	0.48
Windows, averaging, glazing + frame		
Small plane, single glazing, wood or metal frame	5	0.24
Large pane, single glazing, wood or metal frame	8	0.38
Increase for double glazing	2–3	0.10–0.15
Curtain walls, manufactured units	10–15	0.48–0.72
Brick veneer		
4 in., mortar joints	40	1.92
½ in., mastic	10	0.48
Concrete block		
Lightweight, unreinforced—4 in.	20	0.96
6 in.	25	1.20
8 in.	30	1.44
Heavy, reinforced, grouted—6 in.	45	2.15
8 in.	60	2.87
12 in.	85	4.07

in order to simplify and lower the cost of the building construction.

10-2. Dead Load

Dead load consists of the weight of the materials of which the building is constructed: walls, partitions, columns, framing, floors, roofs, and ceilings. In the design of a beam, the dead load must include an allowance for the weight of the beam itself. Table 10-1, which lists the weight of many construction materials, may be used in the computation of dead loads. Dead loads are due to gravity, and they result in downward vertical forces.

10-3. Floor Live Loads

The live load on a floor represents the probable effects created by the occupancy. It includes the weights of human occupants, furniture, equipment, stored materials, and so on. All building codes provide minimum live loads to be used in the design of buildings for various occupancies. Since there is a lack of uniformity among different codes in specifying live loads, the local code should always be used. Table 10-2 contains values for floor live loads as given by the 1982 edition of the *Uniform Building Code.*

Although expressed as uniform loads, code-required values are usually established large enough to account for ordinary concentrations that occur. For offices, parking garages, and some other occupancies, codes often require the consideration of a specified concentrated load as well as the distributed loading. Where buildings are to contain heavy machinery, stored materials, or other contents of unusual weight, these must be provided for individually in the design of the structure.

When structural framing members support large areas, most codes allow some reduction in the total live load to be used for design. The following is the method given in the 1982 edition of the *Uniform Building Code* for determining the reduction permitted for beams, trusses, or columns that support large floor areas.

Except for floors in places of assembly (theaters, etc.), and except for live loads greater than 100 psf [4.79 kN/m^2], the design

TABLE 10-2. Minimum Floor Live Loads

Use or occupancy		Uniform load		Concentrated load	
Description	Description	psf	kN/m²	lb	kN
Armories		150	7.2		
Assembly areas and	Fixed seating areas	50	2.4		
auditoriums and	Movable seating and	100	4.8		
balconies therewith	other areas				
	Stages and enclosed platforms	125	6.0		
Cornices, marquees, and residential balconies		60	2.9		
Exit facilities		100	4.8	Reference a	
Garages	General storage, repair	100	4.8		
	Private pleasure car	50	2.4	Reference a	
Hospitals	Wards and rooms	40	1.9	1000	4.5
Libraries	Reading rooms	60	2.9	1000	4.5
	Stack rooms	125	6.0	1500	6.7
Manufacturing	Light	75	3.6	2000	9.0
	Heavy	125	6.0	3000	13.3
Offices		50	2.4	2000	9.0
Printing plants	Press rooms	150	7.2	2500	11.1
	Composing rooms	100	4.8	2000	9.0
Residential		40	1.9		
Rest rooms		Reference b			
Reviewing stands, grandstands, and bleachers		100	4.8		
Roof decks (occupied)	Same as area served				
Schools	Classrooms	40	1.9	1000	4.5
Sidewalks and driveways	Public access	250	12.0	Reference a	
Storage	Light	125	6.0		
	Heavy	250	12.0		
Stores	Retail	75	3.6	2000	9.0
	Wholesale	100	4.8	3000	13.3

Source: Adapted from the Uniform Building Code, 1982 ed., with permission of the publishers, International Conference of Building Officials.
Wheel loads related to size of vehicles that have access to the area.
Same as the area served or minimum of 50 psf.

live load on a member may be reduced in accordance with the formula

$$R = 0.08(A - 150)$$

$$[R = 0.86(A - 14)]$$

The reduction shall not exceed 40% for horizontal members or for vertical members receiving load from one level only, 60% for other vertical members, nor R as determined by the formula

$$R = 23.1(1 + D/L)$$

In these formulas,

R = reduction in percent
A = area of floor supported by a member
D = unit dead load/ft^2 of supported area
L = unit live load/ft^2 of supported area

In office buildings and certain other building types, partitions may not be permanently fixed in location but may be erected or moved from one position to another in accordance with the requirements of the occupants. In order to provide for this flexibility, it is customary to require an allowance of 15 to 20 psf [0.72 to 0.96 kN/m^2], which is usually added to other dead loads.

10-4. One-Way Solid Slab and Beam Systems

The most widely used and most adaptable poured-in-place concrete floor system is that which utilizes one-way solid slabs supported by one-way spanning beams. This system may be used for single spans, but occurs more frequently with multiple-span slabs and beams in a system such as that shown in Fig. 10-1. In the example shown, the continuous slabs are supported by a series of beams that are spaced at 10 ft center to center. The beams, in turn, are supported by a girder and column system with columns at 30-ft centers, every third beam being supported directly by the columns and the remaining beams being supported by the girders.

Because of the regularity and symmetry of the system shown

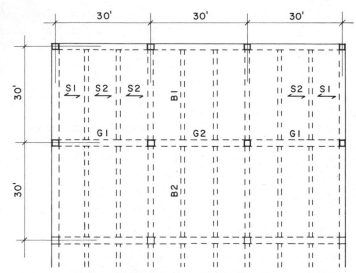

FIGURE 10-1.   Plan of a typical concrete slab-beam-girder framing system.

in Fig. 10-1, there are relatively few different elements in the system, each being repeated several times. While special members must be designed for conditions that occur at the outside edge of the system and at the location of any openings for stairs, elevators, and so on, the general interior portions of the structure may be determined by designing only six basic elements: S1, S2, B1, B2, G1, and G2, as shown in the framing plan. The design of these typical elements is illustrated in Sections 10-5 to 10-7.

In computations for reinforced concrete, the span length of freely supported beams (simple beams) is generally taken as the distance between centers of supports or bearing areas; it should not exceed the clear span plus the depth of beam or slab. The span length for continuous or restrained beams is taken as the clear distance between faces of supports. For a simple beam, that is, a single span having no restraint at the supports, the maximum bending moment for a uniformly distributed load is at the center of the span, and its magnitude is $M = WL/8$. The moment is zero at the supports and is positive over the entire span length. In continuous beams, however, negative bending moments are de-

veloped at the supports and positive moments at or near midspan. This may be readily observed from the exaggerated deformation curve of Fig. 10-2*a*. The exact values of the bending moments depend on several factors, but in the case of approximately equal spans supporting uniform loads, when the live load does not exceed three times the dead load, the bending moment values given in Fig. 10-2 may be used for design.

The values given in Fig. 10-2 are in general agreement with the recommendations of Chapter 8 of the 1977 ACI Code. These values have been adjusted to account for partial live loading of multiple-span beams. Note that these values apply only to uniformly loaded beams. Chapter 8 of the 1977 ACI Code also gives some factors for end-support conditions other than the simple supports shown in Fig. 10-2.

Design moments for continuous-span slabs are given in Fig. 10-3. Where beams are relatively large and the slab spans are small, the rotational (torsional) stiffness of the beam tends to minimize the effect of individual slab spans on the bending in adjacent spans. Thus most slab spans in the slab-beam systems tend to function much like individual spans with fixed ends.

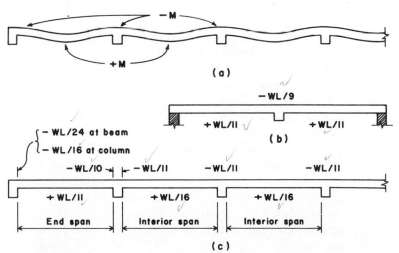

FIGURE 10-2.   Approximate design factors for continuous beams.
(For where L.L ≤ 3 D.L.)

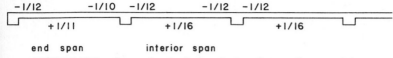

FIGURE 10-3.   Approximate design factors for continuous slabs.

10-5. Design of a Continuous One-Way Solid Slab

The general design procedure for a one-way solid slab was illustrated in Section 9-10. The example given there is for a simple
span slab. The following example illustrates the procedure for the
design of a continuous solid one-way slab.

Example.   A solid one-way slab is to be used for a framing system similar to that shown in Fig. 10-1. Column spacing is 30 ft,
with evenly spaced beams occurring at 10 ft center to center.
Superimposed loads on the structure (floor live load plus other
construction dead load) are a total of 160 psf. Use f'_c = 3 ksi [20.7
MPa] and Grade 40 reinforcement with f_y = 40 ksi [276 MPa] and
f_s = 20 ksi [138 MPa]. Determine the thickness for the slab, and
pick its reinforcement.

Solution: To find the slab thickness, we consider three factors:
the minimum thickness for deflection, the minimum effective
depth for the maximum moment, and the minimum effective
depth for the maximum shear. For all of these we must first
determine the span of the slab. For design purposes this is taken
as the clear span, which is the dimension from face to face of the
supporting beams. With the beams at 10-ft centers, this dimension
is 10 ft, less the width of one beam. Since the beams are not given,
we will assume a dimension for them. In practice we would proceed from the slab design to the beam design, after which the
assumed dimension could be verified. For this example we will
assume a beam width of 12 in., yielding a clear span of 9 ft.

We consider first the minimum thickness required for deflection. If the slabs in all spans have the same thickness (which is the
most common practice), the critical slab is the end span, since
there is no continuity of the slab beyond the end beam. While the

beam will offer some restraint, it is best to consider this as a simple support; thus we use the factor of $L/30$ from Table 9-2.

$$\text{minimum } t = \frac{L}{30} = \frac{9 \times 12}{30} = 3.6 \text{ in.}$$

We will try a 4-in.-thick slab, for which the dead weight of the slab will be

$$w = \frac{4}{12} \times 150 = 50 \text{ lb/ft}^2$$

and the total design loading will thus be $50 + 160 = 210$ lb/ft^2.

We next consider the maximum bending moment. Inspection of the moment values given for various locations in Fig. 10-3 shows the maximum value to be $\frac{1}{10} wL^2$. With the span and loading as determined, the maximum moment is thus

$$M = \frac{1}{10} wL^2 = \frac{1}{10} (210)(9)^2 = 1701 \text{ ft-lb}$$

This moment should now be compared to the balanced moment capacity for the design section, using the relationships as discussed for rectangular beams in Section 6-4. For this computation we must assume an effective depth for the design section. This dimension will be the slab thickness minus the concrete cover and one-half the bar diameter. With the reinforcing not yet determined, we will assume an effective depth equal to the slab thickness minus 1.125 in., which will be exactly true with the usual cover of $\frac{3}{4}$-in. and a $\frac{3}{4}$ in.-diameter (No. 6) bar. Then using the balanced R factor from Table 6-2, the maximum resisting moment for the 12-in.-wide design section is

$$M_R = Rbd^2 = 0.226(12)(2.875)^2 = 22.416 \text{ k-in.}$$

or,

$$M_R = 22.416 \times \frac{1000}{12} = 1868 \text{ ft-lb}$$

As this value is in excess of the required maximum moment, the slab will be adequate for flexural stress.

It is not practicable to use shear reinforcement in one-way slabs, and consequently the maximum unit shear stress must be kept within the limit for the concrete alone. The usual procedure is to check the shear stress with the effective depth determined for bending before proceeding to find A_s. Except for very short span slabs with excessively heavy loadings, shear stress is seldom critical.

Finally, before proceeding with the design of the reinforcing, we should verify our slab thickness for shear stress. For an interior span, the maximum shear will be $wL/2$, but for the end span it is the usual practice to consider some unbalanced condition for the shear due to the discontinuous end. We therefore use a maximum shear of 1.15 $wL/2$, or an increase of 15% over the simple beam shear value. Thus

$$\text{maximum shear} = V = 1.15 \frac{wL}{2} = 1.15 \times \frac{210 \times 9}{2} = 1087 \text{ lb}$$

$$\text{and maximum shear stress} = v = \frac{V}{bd} = \frac{1087}{12 \times 2.875} = 31.5 \text{ psi}$$

This is considerably less than the limit for the concrete alone ($v_c = 1.1\sqrt{f'_c} = 60$ psi), so the assumed slab thickness is not critical for shear stress.

Having thus verified our choice for the slab thickness, we may now proceed with the design of the reinforcing. For a balanced section, Table 6-2 yields a value of 0.872 for the j factor. However, since all of our reinforced sections will be classified as under-reinforced (actual moment less than the balanced limit), we will use a slightly higher value, say 0.90, for j in the design of the reinforcing.

Referring to Fig. 10-4, we note that there are five critical locations for which a moment must be determined and the required steel area computed. Reinforcing in the top of the slab must be computed for the negative moments at the end support, at the first interior beam, and at the typical interior beam. Reinforcing in the bottom of the slab must be computed for the positive moments at midspan in the first span and the typical interior spans. The de-

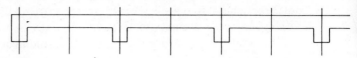

Moment factor:

| $-1/12$ | $+1/11$ | $-1/10$ | $+1/16$ | $-1/12$ | $+1/16$ |

Required A_s - in²:

| 0.329 | 0.359 | 0.394 | 0.247 | 0.329 | 0.247* |

Required spacing of bars - in. c/c :   (Table 15-2)

No. 3 at	4	3.75	3.25	5.75	4	5.75
No. 4 at	7	6.5	6	9.5	7	9.5
No. 5 at	11	10	9.5	15	11	15

Selection:

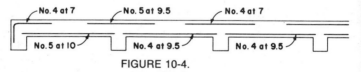

FIGURE 10-4.

sign for these conditions is summarized in Fig. 10-4. For the data displayed in the figure we note the following:

maximum spacing of reinforcing = $3 \times t = 3 \times 4 = 12$ in.

design moment = M = (moment factor F) $\times wL^2$

$$= F \times (210)(9)^2 \times 12$$

$$= F \times 204{,}120 \qquad \text{(in.-lb units)}$$

$$\text{required } A_s = \frac{M}{f_s jd} = \frac{F \times 204{,}120}{(20{,}000)(0.9)(2.875)} = F \times 3.944$$

Using data from Table 9-4, Fig. 10-4 shows required spacings for No. 3, 4, and 5 bars. A possible choice for the slab reinforcing, using all straight bars, is shown at the bottom of the figure.

Note: In the design of the slab in this section, we have used the working stress method. For strength design, the procedure is essentially the same, except for the use of the factored load and the proper formulas for stress analysis. For brevity, we will therefore not illustrate the complete solution by the strength design method. This applies also to the examples for the beam and girder designs in the following sections.

ultimate Design (handwritten annotation)

Problem 10-5-A.   A solid one-way slab is to be used for a framing system similar to that shown in Fig. 10-1. Column spacing is 36 ft, with regularly spaced beams occurring at 12 ft center to center. Superimposed loads on the structure are a total of 180 psf [8.62 kN/m²]. Use f'_c = 3 ksi [20.7 MPa] and Grade 40 reinforcing with f_y = 40 ksi [276 MPa] and f_s = 20 ksi [138 MPa]. Determine the thickness for the slab, and select the size and spacing for the bars.

10-6.   Design of a Continuous Beam

The following example illustrates the design procedure for a continuous reinforced concrete beam subjected to a uniformly distributed loading.

Example.   Design the continuous beam (B1-B2) for the framing system shown in Fig. 10-1. Use the data given for the slab design example in Section 10-5.

Solution: Assuming the girders to be approximately 18 in. wide, the clear spans for the beams will be as shown at (*a*) in Fig. 10-5. The loading for the beam is determined as follows.

From the slab design (see Section 10-5):

Assumes live load = 100 psf (handwritten annotation)

$$\text{superimposed dead load} = 60 \text{ psf}$$

$$\text{slab weight} = 4 \times \frac{150}{12} = 50 \text{ psf}$$

For the reduced live load:

$$\text{area supported by the beam} = 10 \times 28.5 = 285 \text{ ft}^2$$

Then, from Section 10-3, the reduction is

$$R = 0.08(A - 150) = 0.08(285 - 150) = 10.8\% \text{ (say 10)}$$

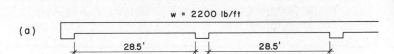

(a)

w = 2200 lb/ft

|← 28.5' →| |← 28.5' →|

(b) Moment Factors: (See Fig. 10-2)

$$C = \frac{1}{24} \qquad \frac{1}{11} \qquad \frac{1}{10} \; \frac{1}{11} \qquad \frac{1}{16}$$

(c) Required $A_s = CwL^2$: (in.^{2})

Top: 1.99   4.78

Bottom: 3.98   2.74

(d) Bar Selection:

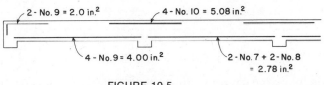

2 - No. 9 = 2.0 in.^{2} 4 - No. 10 = 5.08 in.^{2}

4 - No. 9 = 4.00 in.^{2} 2 - No. 7 + 2 - No. 8 = 2.78 in.^{2}

FIGURE 10-5.

Use 90% of live load, or 90 psf

Beam weight: Assume at 200 lb/ft of span

Then the total unit uniformly distributed load on the beam is

$$w = 10(60 + 50 + 90) + 200 = 2200 \text{ lb/ft}$$

This is the loading shown in Fig. 10-5.

The design moments for the various critical locations along the beam span are determined using the moment factors shown at (b) in Fig. 10-5. These are obtained from Fig. 10-2.

We will assume the use of the typical beam section shown in Fig. 10-6, consisting of a rectangular shape with single No. 3 U-stirrups and 1.5 in. of concrete cover on the reinforcing. Where the beams and girders meet in plan, the possible interference of the intersecting reinforcing bars must be considered. As shown in Fig. 10-6, we assume the tops of the beams and girders to be the same, but the bottoms to be at different levels, due to a larger girder depth. Thus the bottom bars in the members will be at different levels, permitting the optimal placing of each set—as-

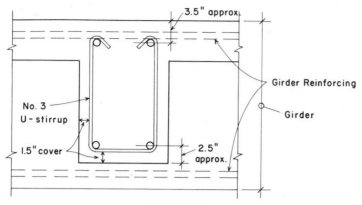

FIGURE 10-6.

sumed for design to be 2.5 in. from the center of the bars to the bottom surface of the concrete. The top bars, if optimally placed in both members, would occur at the same level, making it impossible to extend the beam bars through the girders. We therefore concede the prime location to the heavier loaded girder and assume the top bars in the beams to be slightly farther from the top surface—assumed for design to be 3.5 in., as shown in Fig. 10-6.

The problem of locating the top bars is most critical at the columns, where the negative moments are a maximum in both the beams and the girders. If there are no continuous top bars in the grider, this problem may not exist for the beams that occur between the columns (those actually supported by the girders), which may indicate the desirability of designing a special beam for the one that is directly supported by the columns.

The next step in the beam design procedure is to select the dimensions for the concrete section. If compressive reinforcing (Section 9-11) is not desired, the limits for the dimensions are those required for a balanced section. To determine these, we use the following data from Table 6-2.

$$R = 0.226, \quad j = 0.872, \quad p = 0.0129$$

The largest negative moment occurs at the outside face of the

interior column where the moment factor is $\frac{1}{10}$. We thus compute the maximum negative moment as

$$M = \frac{1}{10} wL^2 = \frac{1}{10} (2.200)(28.5)^2 = 179 \text{ k-ft}$$

Then, using the relationship $M = Rbd^2$, we find

$$\text{required } bd^2 = \frac{M}{R} = \frac{179 \times 12}{0.226} = 9504$$

We may now try various combinations of b and d.

If $b = 12$ in.,

$$d = \sqrt{\frac{9504}{12}} = 28.14 \text{ in.}$$

If $b = 14$ in.,

$$d = \sqrt{\frac{9504}{14}} = 26.05 \text{ in.}$$

And so on.

The selection of the beam dimensions will be influenced by other considerations, including design for shear. For an approximate design, we may consider the beam to be able to carry a maximum computed shear stress of twice that of which the concrete alone is capable. (See Sectin 7-3.) Thus if $b = 12$ in. and $d = 28$ in. and $v_c = 1.1\sqrt{f'_c} = 1.1\sqrt{3000} = 60$ psi, then the maximum permitted end shear force is

$$V = 2v_c bd = 2 \times 60 \times 12 \times 28 = 40{,}320 \text{ lb}$$

This shear value may be compared to the maximum design shear for the beam, assumed for design to be 15% greater than the simple beam shear. Thus

$$V = 1.15 \times \frac{wL}{2} = 1.15 \times \frac{2200 \times 28.5}{2} = 36{,}053 \text{ lb}$$

This indicates that a moderate amount of shear reinforcement will be required if a balanced section is used. (Remember that

some shear reinforcement is usually desirable, as discussed in Chapter 7.)

Unless deflections are to be computed, the beam should have an overall height not less than that given in Table 9-2. For the end span of our beam, with one end discontinuous, the clear span of 28.5 ft requires a minimum beam height of

$$h = \frac{28.5 \times 12}{23} = 14.9 \text{ in.}$$

Let us try a section with dimensions of $b = 14$ in. and $h = 29$ in. Then the effective depths for design will be as follows.

For negative moment:

$$29 - 3.5 = 25.5 \text{ in.}$$

For positive moment:

$$29 - 2.5 = 26.5 \text{ in.}$$

The areas required for steel may thus be computed as follows:

For $-M$:

$$A_s = \frac{M}{f_s jd} = \frac{CwL^2}{f_s jd} = \frac{C(2.200)(28.5)^2(12)}{(20)(0.88)(25.5)}$$

$$= 47.8C$$

For $+M$:

$$A_s = \frac{M}{f_s(d - t/2)} = \frac{C(2.200)(28.5)^2(12)}{(20)(26.5 - 2)}$$

$$= 43.8C$$

in which C is the moment factor (Fig. 10-5b).

Using these expressions, the required steel areas are determined as shown in Fig. 10-5c. Using all straight bars, except for the hooks at the spandrel beam, a possible selection for the bars is shown in Fig. 10-5d.

Table 9-1 may be used to verify that the 14-in.-wide beam can contain the bars with adequate spacing. For the worst case—that

of the four No. 10 bars—the table indicates a minimum width of 13 in. The other cases will be less critical, so the beam has adequate width for this purpose.

There are two shear designs that must be done for this beam. The first occurs at the outside end of the exterior span and at both ends of the interior span. At these locations, the maximum shear is $wL/2$. At the inside end of the exterior span, the ACI Code requires a design shear 15% greater than the simple beam shear.

Let us first consider the lower shear value. For this condition, we compute

$$V = \frac{wL}{2} = \frac{(2200)(28.5)}{2} = 31,350 \text{ lb}$$

$$v = \frac{V}{bd} = \frac{31,350}{14 \times 25.5} = 88 \text{ psi}$$

Using the procedure described in Section 7-5, the data pertaining to the design of the shear reinforcement is presented in Fig. 10-7. The stirrup selection shown results in a total of nine stirrups at the beam end for this case.

Figure 10-8 presents data pertaining to the shear design at the inside ends of the exterior spans. The shear stress diagram for

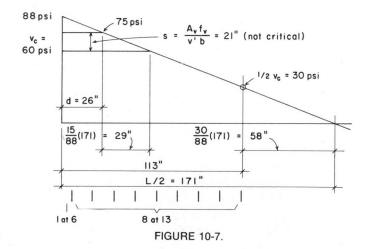

FIGURE 10-7.

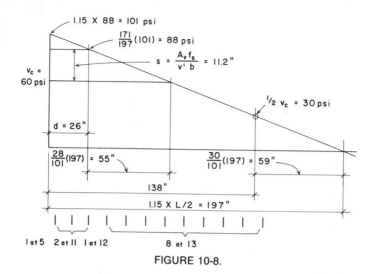

FIGURE 10-8.

this case has been modified by increasing both the maximum shear stress and the total horizontal length by 15%. The stirrup selection shown results in two additional stirrups and a slightly closer spacing near the support.

A final consideration for the beam is that of the development lengths for the bars. If the bars are placed as shown in Fig. 10-5d, with no partial length bars, the only two concerns are for the cutoff points for the top bars and the anchorage of the top bars at the spandrel girder.

Cutoff points are often established by rules of thumb for common situations. The recommendations given in Fig. 10-9 are typical of those used in practice. Where ordinary conditions exist, these approximations are usually adequate. Situations where computations are indicated include the following:

Live load exceeds 1.5 times dead load.

Adjacent beam spans differ by more than 20%.

Reinforcement consists of Grade 60 bars of large diameter (Nos. 9 to 11).

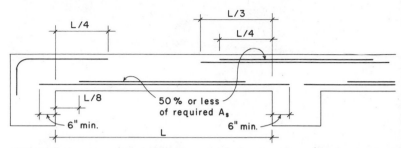

FIGURE 10-9. Example of typical minimum requirements used for development and cutoff lengths.

For the anchorage at the spandrel girder, Table 8-2 indicates that a standard hook will provide an equivalent embedment length of 14.4 in. for the No. 9 bar. Table 8-1 requires a total development length of 41 in. for $f'_c = 3$ ksi and $f_y = 40$ ksi. Thus, in theory, the spandrel girder must provide for embedment of $41 - 14.4 = 26.6$ in. in addition to the hook (distance L_1 as shown in Fig. 10-10). With the required cover for the bars, this would require a minimum width of about 29 in. for the girder. If the girder is not this wide, it will be necessary to provide more total area of steel to reduce the tension stress, or to use smaller bars that require less development length. For example, if we change this reinforcement to four No. 7 bars, which provide a total of 2.40 in.^{2}, the tension stress will be reduced by approximately 20%. This reduction in tension stress, plus the shorter lengths required for development of the smaller bars, will make it possible for the girder to be as narrow as 16 in.

Another concern for the hooked bar at the spandrel is the proper development of the length shown as L_2 in Fig. 10-10. From data in Figs. 5-1 and 5-2, it may be found that this distance must be approximately 14 in. for the formation of a proper hook for a No. 7 bar. With the proper concrete cover, this would require a minimum overall girder height of 17 in.—not a critical condition for our example.

Problem 10-6-A. Using data from Problem 10-5-A, design the three-span continuous beam.

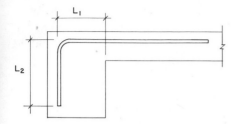

FIGURE 10-10.

10-7. Design of a Continuous Girder

The following example illustrates an approximate design procedure for a continuous girder subjected to a combined loading
consisting of both uniformly distributed and concentrated loads.

Example.   Design the three-span continuous girder (G1-G2-G1)
for the framing system shown in Fig. 10-5. Use the necessary data
from the examples in Sections 10-5 and 10-6.
Solution: We first determine the loadings. Since every third
beam is carried directly by the columns, the girders carry approximately two-thirds of the full column bay area, or 600 ft.[2] On this
basis, the live load reduction will be (see Section 10-3):

$$R = 0.08(A - 150) = 0.08(600 - 150) = 36\%$$

and the design live load will be 64 psf.
  The concentrated loads represent the end reactions of the
beams. For the interior girder (not at the edge of the building), the
end reactions from the beams on each side of the girder add up to
the full load on one beam. Therefore, we determine the concentrated load as follows.

Live load:	64 psf × 10 ft × 28.5 ft =		18,240 lb
Dead load:	110 psf × 10 × 28.5 =		31,350 lb
Beam weight:	24 × 12 × $\frac{150}{144}$ × 28.5 =		8,550 lb
Total load:			58,140 lb

For simplicity, we will round this off to 58 k.
  The uniformly distributed load consists of the weight of the

girder plus the superimposed loads on the strip of floor immediately over the girder. (The 18-in.-wide strip is not included in the beam load.) Assuming a girder section that is 18 in. wide and 30 in. in overall height, the unit distributed load is

Girder weight: $18 \times 30 \times \frac{150}{144} =$ 563 lb/ft

Superimposed load: 124 psf $\times$ 1.5 ft = 186 lb/ft

Total unit load: 749 lb/ft

For simplicity, we will round this off to 0.75 k/ft.

Note that the superimposed load does not include the slab weight, since we have included this by using the total girder section dimensions. This loading is applied to the girder as shown in Fig. 10-11. For an approximate design, we will consider the total load to be applied as a single distributed load and proceed to use the moment factors from the ACI Code, as illustrated for the beam design in Section 10-6.

The moment factors, taken from Fig. 10-2, are shown immediately below the figure in Fig. 10-11. Below these are shown the moments for the critical points along the beam span, as determined by using the total load, as described previously. We now consider the choice for the girder dimensions, based on considerations of flexure, shear, and deflection.

For flexure, we will most likely consider the use of a section with some compressive reinforcing for the maximum negative moment at the interior column. This will permit some reduction in the girder size (below that required for a balanced section by the working stress method) and will enhance the strength of the girder-column bent (rigid vertical frame) where this is to be used for resistance to wind or seismic forces on the building. For a reasonable limit, we recommend a section whose balanced moment capacity is approximately two-thirds that of the total required resisting moment. Thus

$$\text{design } M = \frac{2}{3} \times 393 = 262 \text{ k-ft}$$

$$\text{required } bd^2 = \frac{M}{R} = \frac{262 \times 12}{0.226} = 13{,}912 \text{ in.}^3$$

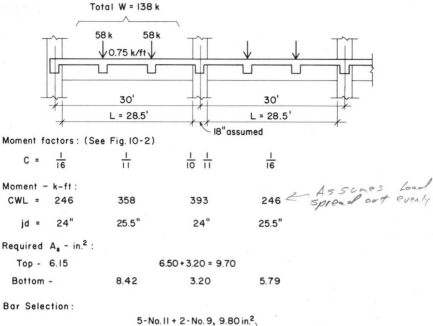

Moment factors: (See Fig. 10-2)

$C = \dfrac{1}{16}$ $\dfrac{1}{11}$ $\dfrac{1}{10}$ $\dfrac{1}{11}$ $\dfrac{1}{16}$

Moment − k-ft:

CWL =	246	358	393	246
jd =	24"	25.5"	24"	25.5"

← Assumes load spread out evenly

Required A_s − in.² :

Top − 6.15   6.50 + 3.20 = 9.70

Bottom −   8.42   3.20   5.79

Bar Selection:

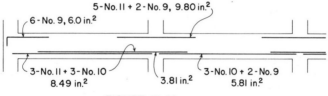

FIGURE 10-11.

and if $b = 18$ in.,

$$d = \sqrt{\frac{13{,}912}{18}} = 27.8 \text{ in.}$$

This indicates that the assumed section of 18 in. by 30 in. is adequate for flexure.

Before proceeding with the design of the reinforcing, it is wise to check the shear condition, to be sure that the section dimensions are reasonable. For this loading, the shear is almost constant for the outer thirds of the spans—from the supports to the

beams. Thus the stirrups will be spaced at a relatively constant dimension throughout these distances. If the required spacing is very close, this could mean that a very large number of stirrups would be required. It is probably wise, therefore, to pick a stirrup size and the beam dimensions to permit a reasonable stirrups spacing, say 6 in. or more.

As with the beam in Section 10-6, there are two shear conditions to consider: that at the interior ends of the exterior spans and that at other locations. Let us use the critical condition at the interior ends of the exterior spans, where we may use a design shear of 1.15 times the simple beam shear. This will result in a maximum shear force of

$$V = 1.15 \times \frac{138}{2} = 79.35 \text{ k}$$

Since this will be reduced slightly for the critical design at a d distance from the support, we may choose a maximum shear value of approximately 78 k for design. Thus the maximum shear stress for the stirrup spacing is

$$v = \frac{V}{bd} = \frac{78,000}{18 \times 27.5} = 158 \text{ psi}$$

working Stress

With an allowable stress of 60 psi $(1.1 \sqrt{f_c'})$ this leaves a stress of 98 psi to be taken by the stirrups. If we use a No. 4 U-stirrup, the required spacing is

$$s = \frac{A_v f_v}{v' b} = \frac{2(0.20) \times 20,000}{98 \times 18} = 4.54 \text{ in.}$$

This would require an average spacing of about 5 in. for the stirrups in the end thirds of the girder spans. While not impossible, it is questionable from an economic viewpoint. We will, however, stick with the beam dimensions for the time.

Data relating to the design of the flexural reinforcing for the girder is displayed at the bottom of Fig. 10-11. The area requirements have been determined by the usual procedures for the working stress method. We note the following for the various critical moment locations.

1. At the outside column: The moment can be sustained without compressive reinforcing; j is approximately that for a balanced section. Bars should be as small in diameter as possible to reduce development problems at the column.
2. At the inside columns: Area requirement for the top bars indicates the $A_{s1} + A_{s2}$ values, as described in Section 9-11. Bottom bars are compressive reinforcing and must be developed by extension through the column and into the interior span.

The procedure for the design of web reinforcing for shear for the girder is essentially similar to that for the beam. As determined previously, a considerable number of closely spaced stirrups will be required for the beam size chosen. If other factors favor it, it may be wise to increase the girder width and/or depth in order to reduce the amount of shear reinforcing. It should also be noted that the stirrups must function as ties for the compressive reinforcing at the interior column.

Problem 10-7-A.   Using data from Problem 10-5-A, design the continuous girder.

10-8. Concrete One-Way Joist Construction

Figure 10-12 shows a partial framing plan and some details for a type of construction that utilizes a series of very closely spaced beams and a relatively thin solid slab. Because of its resemblance to ordinary wood joist construction, this is called concrete joist construction. This system is generally the lightest (in dead weight) of any type of flat-spanning, poured-in-place concrete construction and is structurally well suited to the light loads and medium spans of office buildings and commercial retail buildings.

Slabs as thin as 2 in. and joists as narrow as 4 in. are used with this construction. Because of the thinness of the parts and the small amount of cover provided for reinforcement (typically ¾ to 1 in. for joists versus 1.5 in for ordinary beams), the construction has very low resistance to fire, especially when exposed from the underside. It is therefore necessary to provide some form of fire protection, as for steel construction, or to restrict its use to situations where high fire ratings are not required.

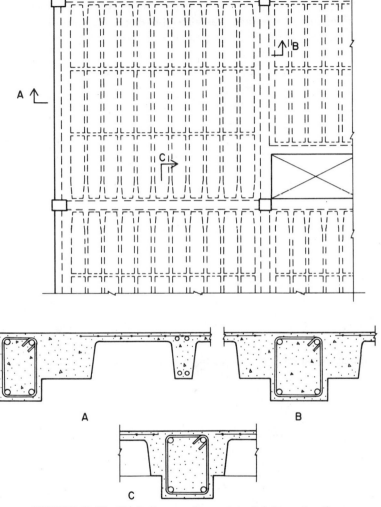

FIGURE 10-12. Typical concrete one-way joist construction.

The relatively thin, short span slabs are typically reinforced with welded wire mesh rather than ordinary deformed bars. Joists are often tapered at their ends, as shown in the framing plan in Fig. 10-12. This is done to provide a larger cross section for increased resistance to shear and negative moment at the supports. Shear reinforcement in the form of single vertical bars may be provided, but is not frequently used.

Early joist construction was produced by using lightweight hollow clay tile blocks to form the voids between joists. These blocks were simply arranged in spaced rows on top of the forms, the joists being formed by the spaces between the rows. The resulting construction provided a flat underside to which a plastered ceiling surface could be directly applied. Hollow, lightweight concrete blocks later replaced the clay tile blocks. Other forming systems have utilized plastic-coated cardboard boxes, fiber glass reinforced pans, and formed sheet metal pans. The latter method was very widely used, the metal pans being pried off after the pouring of the concrete and reused for several additional pours. The tapered joist cross section shown in Fig. 10-12 is typical of this construction, since the removal of the metal pans requires it.

Wider joists can be formed by simply increasing the space between forms, with large beams being formed in a similar manner or by the usual method of extending a beam stem below the construction, as shown for the beams in Fig. 10-12. Because of the narrow joist forms, cross bridging is usually required, just as with wood joist construction. The framing plan in Fig. 10-12 shows the use of two bridging strips in the typical bay of the framing.

Design of joist construction is essentially the same as for ordinary slab and beam construction. Some special regulations are given in the ACI Code for this construction, such as the reduced cover mentioned previously. Because joists are so commonly formed with standard-sized metal forms, there are tabulated designs for typical systems in various handbooks. The *CRSI Handbook* (Ref. 3) has extensive tables offering complete designs for various spans, loadings, pan sizes, and so on. Whether for final

design or simply for a quick preliminary design, the use of such tables is quite efficient.

One-way joist construction was highly popular in earlier times, but has become less utilized, due to its lack of fire resistance and the emergence of other systems. The popularity of lighter, less fire resistive ceiling construction has been a contributing factor. In the right situation, however, it is still a highly efficient type of construction.

10-9. Concrete Waffle Construction

Waffle construction consists of two-way spanning joists that are formed in a manner similar to that for one-way spanning joists, using forming units of metal, plastic, or cardboard to produce the void spaces between the joists. The most widely used type of waffle construction is the waffle flat slab, in which solid portions around column supports are produced by omitting the void-making forms. An example of a portion of such a system is shown in Fig. 10-13. This type of system is analogous to the solid flat slab which will be discussed in Section 10-10. At points of discontinuity in the plan—such as at large openings or at edges of the building—it is usually necessary to form beams. These beams may be produced as projections below the waffle, as shown in Fig. 10-13, or may be created within the waffle depth by omitting a row of the void-making forms, as shown in Fig. 10-14.

If beams are provided on all of the column lines, as shown in Fig. 10-14, the construction is analogous to the two-way solid slab with edge supports, as discussed in Section 10-10. With this system, the solid portions around the column are not required, since the waffle itself does not achieve the transfer of high shear or development of the high negative moments at the columns.

As with the one-way joist construction, fire ratings are low for ordinary waffle construction. The system is best suited for situations involving relatively light loads, medium to long spans, approximately square column bays, and a reasonable number of multiple bays in each direction.

For the waffle construction shown in Fig. 10-13, the edge of the structure represents a major discontinuity when the column sup-

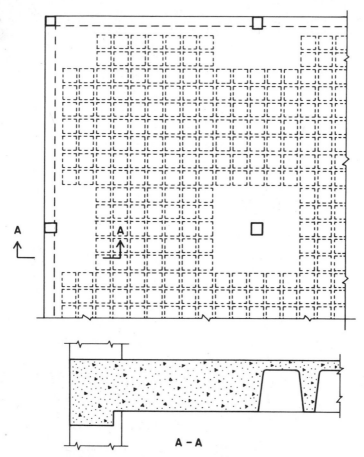

FIGURE 10-13. Typical concrete waffle construction.

ports occur immediately at the edge, as shown. Where planning permits, a more efficient use of the system is represented by the partial framing plan shown in Fig. 10-15, in which the edge occurs some distance past the columns. This projected edge provides a greater shear periphery around the column and helps to generate a negative moment, preserving the continuous character of the spanning structure. With the use of the projected edge, it may be

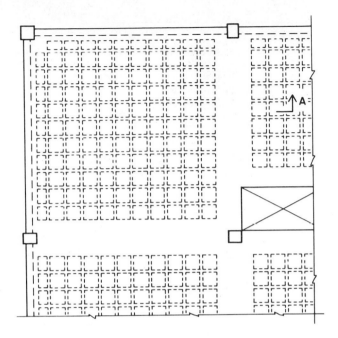

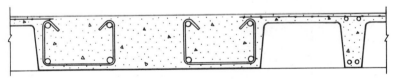

FIGURE 10-14. Typical concrete waffle construction with coulum-line beams within the waffle depth.

possible to eliminate the edge beams shown in Fig. 10-13, thus preserving the waffle depth as a constant.

Another variation for the waffle is the blending of some one-way joist construction with the two-way waffle joists. This may be achieved by keeping the forming the same as for the rest of the waffle construction and merely using the ribs in one direction to create the spanning structure. One reason for doing this would be a situation similar to that shown in Fig. 10-14, where the large

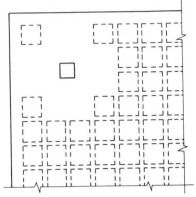

FIGURE 10-15.   Plan of waffle construction with cantilevered edge and no edge beams.

opening for a stair or elevator results in a portion of the waffle (the remainder of the bay containing the opening) being considerably out of square, that is, having one span considerably greater than the other. The joists in the short direction in this case will tend to carry most of the load due to their greater stiffness (less deflection than the longer spanning joists that intersect them). Thus the short joists would be designed as one-way spanning members and the longer joists would have only minimum reinforcing and serve as bridging elements.

The two-way spanning waffle systems are quite complex in structural behavior and their investigation and design is beyond the scope of this book. Some aspects of this work are discussed in the next article, since there are many similarities between the two-way spanning waffle systems and the two-way spanning solid slab systems. As with the one-way joist system, there are some tabulated designs in various handbooks that may be useful for either final or preliminary design. The *CRSI Handbook* (Ref. 3) mentioned previously has some such tables.

10-10. Two-Way-Spanning Solid Slab Construction

If reinforced in both directions, the solid concrete slab may span two ways as well as one. The widest use of such a slab is in flat

slab or flat plate construction. In flat slab construction, beams are used only at points of discontinuity, with the typical system consisting only of the slab and the strengthening elements used at column supports. Typical details for a flat slab system are shown in Fig. 10-16. Drop panels consisting of thickened portions square in plan are used to give additional resistance to the high shear and negative moment that develops at the column supports. Enlarged portions are also sometimes provided at the tops of the columns (called column capitals) to further reduce the stresses in the slab.

Two-way slab construction consists of multiple bays of solid

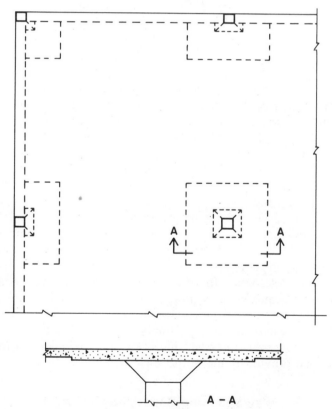

FIGURE 10-16.   Typical concrete flat slab construction with drop panels and column caps.

two-way-spanning slabs with edge supports consisting of bearing walls of concrete or masonry or of column-line beams formed in the usual manner. Typical details for such a system are shown in Fig. 10-17.

Two-way solid slab construction is generally favored over waffle construction where higher fire rating is required for the unprotected structure or where spans are short and loadings high. As with all types of two-way spanning systems, they function most efficiently where the spans in each direction are approximately the same.

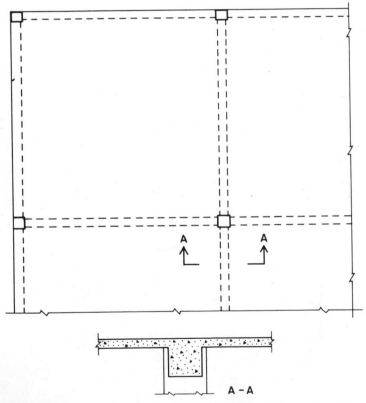

FIGURE 10-17. Typical two-way spanning concrete slab construction with edge supports.

For investigation and design, the flat slab (Fig. 10-16) is considered to consist of a series of one-way-spanning solid slab strips. Each of these strips spans through multiple bays in the manner of a continuous beam and is supported either by columns or by the strips that span in a direction perpendicular to it. The analogy for this is shown in Fig. 10-18a.

As shown in Fig. 10-18b, the slab strips are divided into two types: those passing over the columns, and those passing be-

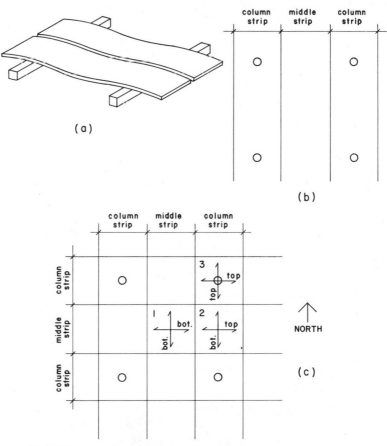

FIGURE 10-18.   Development of the two-way concrete flat slab.

tween columns—called middle strips. The complete structure consists of the intersecting series of these strips, as shown in Fig. 10-18c. For the flexural action of the system there is two-way reinforcing in the slab at each of the boxes defined by the intersections of the strips. In Box 1 in Fig. 10-18c, both sets of bars are in the bottom portion of the slab, due to the positive moment in both intersecting strips. In Box 2, the middle-strip bars are in the top (for negative moment) while the column-strip bars are in the bottom (for positive moment). And, in Box 3, the bars are in the top in both directions.

10-11. Use of Design Aids

The design of various elements of reinforced concrete can be aided—or in many cases totally achieved—by the use of various prepared materials. Handbooks (see Refs. 3, 9, and 10) present complete data for various elements, such as footings, columns, one-way slabs, joist construction, waffle systems, and two-way slab systems. For the design of a single footing or a one-way slab, the handbook merely represents a convenience, or a shortcut, to a final design. For columns subjected to bending, for waffle construction, and for two-way slab systems, "longhand" design (without aid other than a pocket calculator) is really not feasible. In the latter cases, handbook data may be used to establish a reasonable preliminary design, which may then be custom fit to the specific conditions by some investigation and computations. Even the largest of handbooks cannot present all possible combinations of values of f'_c, grade of reinforcing bars, value of superimposed loads, and so on. Thus only coincidentally will handbook data be exactly correct for any specific design job.

In the age of the computer, there is a considerable array of software available for the routine tasks of structural design. For many of the complex and laborious problems of design of reinforced concrete structures, these are a real boon for anyone able to utilize them.

11

Reinforced
Concrete Columns
||||||||||||||||||||||||||||||||||||||||||||||||||

11-1. Introduction

The practicing structural designer customarily uses tables or a computer-aided procedure to determine the dimensions and reinforcing for concrete columns. The complexity of analytical formulas and the large number of variables make it impractical to perform design for a large number of columns solely by hand computation. The provisions relating to the design of columns in the 1977 ACI Code are quite different from those of the working stress design method in the 1963 Code. The current code does not permit design of columns by the working stress method, but it rather requires that the service load capacity of columns be determined as 40% of that computed by strength design procedures.

$\dfrac{1}{1.7}$

.7

11-2. Columns with Axial Load plus Bending Moment

Due to the nature of most concrete structures, current design practices generally do not consider the possibility of a concrete column with axial compression alone. That is to say, the existence of some bending moment is always considered together

182

with the axial force. Figure 11-1 illustrates the nature of the so-called *interaction response* for a concrete column, with a range of combinations of axial load plus bending moment. In general, there are three basic ranges of this behavior, as follows:

1. Large Axial Force, Minor Moment. For this case the moment has little effect, and the resistance to pure axial force is only negligibly reduced.
2. Significant Values for Both Axial Force and Moment. For this case the analysis for design must include the full combined force effects, that is, the interaction of the axial force and the bending moment.
3. Large Bending Moment, Minor Axial Force. For this case, the column behaves essentially as a doubly reinforced (tension and compression reinforced) member, with its capacity for moment resistance affected only slightly by the axial force.

In Fig. 11-1 the solid line on the graph represents the true response of the column—a form of behavior verified by many load tests on laboratory specimens. The dashed line figure on the graph represents the generalization of the three types of response just described.

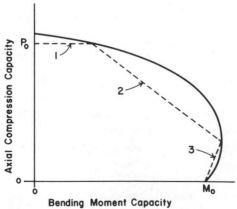

FIGURE 11-1. Interaction of axial compression and bending moment for a reinforced concrete column.

The terminal points of the interaction response—pure axial compression or pure bending moment—may be reasonably easily determined. The interaction responses between these two limits require complex analyses, beyond the scope of this book.

11-3. Types of Reinforced Concrete Columns

Reinforced concrete columns for buildings generally fall into one of the following categories:

1. Square tied columns.
2. Round spiral columns.
3. Rectangular tied columns.
4. Columns of other geometries (hexagonal, L-shaped, T-shaped, etc.) with either ties or spirals.

In tied columns the longitudinal reinforcing is held in place by loop ties made of small-diameter reinforcing bars, commonly No. 3 or No. 4. Such a column is represented by the square section shown in Fig. 11-2a. This type of reinforcing can quite readily accommodate other geometries as well as the square. The design of such a column is discussed in Section 11-4.

Spiral columns are those in which the longitudinal reinforcing is placed in a circle, with the whole group of bars enclosed by a continuous cylindrical spiral made from steel rod or large-diameter steel wire. Although this reinforcing system obviously works best with a round column section, it can be used also with other geometries. A round column of this type is shown in Fig. 11-2b.

Experience has shown the spiral column to be slightly stronger than an equivalent tied column with the same amount of concrete and reinforcing. For this reason, code provisions allow slightly more load on spiral columns. Spiral reinforcing tends to be expensive, however, and the round bar pattern does not always mesh well with other construction details in buildings. Thus tied columns are often favored where restrictions on the outer dimensions of the sections are not severe.

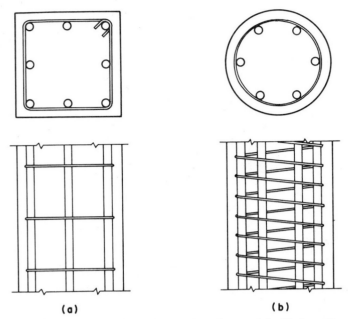

FIGURE 11-2. Typical reinforced concrete columns: (a) with loop ties and (b) with a spiral wrap.

11-4. General Requirements for Reinforced Concrete Columns

Code provisions and practical construction considerations place a number of restrictions on column dimensions and choice of reinforcing.

Column Size. The current code does not contain limits for column dimensions. For practical reasons, the following limits are recommended. Rectangular tied columns should be limited to a minimum area of 100 in.^{2} and a side dimension of 10 in. if square and 8 in. if oblong. Spiral columns should be limited to a minimum size of 12 in. if either round or square.

Reinforcing. Minimum bar size is No. 5. The minimum number of bars is four for tied columns, five for spiral columns. The

minimum amount of area of steel is 1% of the gross column area. A maximum area of steel of 8% of the gross area is permitted, but bar spacing limitations makes this difficult to achieve; 4% is a more practical limit. Section 10.8.4 of the 1977 ACI Code stipulates that for a compression member with a larger cross section than required by considerations of loading, a reduced effective area not less than one-half the total area may be used to determine minimum reinforcement and design strength.

Ties.   Ties shall be at least No. 3 for bars No. 10 and smaller. No. 4 ties should be used for bars that are No. 11 and larger. Vertical spacing of ties shall be not more than 16 times the bar diameter, 48 times the tie diameter, or the least dimension of the column. Ties shall be arranged so that every corner and alternate longitudinal bar is held by the corner of a tie with an included angle of not greater than 135°, and no bar shall be farther than 6 in. clear from such a supported bar. Complete circular ties may be used for bars placed in a circular pattern.

Concrete Cover.   A minimum of 1.5 in. is needed when the column surface is not exposed to weather or in contact with the ground; 2 in. should be used for formed surfaces exposed to the weather or in contact with ground; 3 in. are necessary if the concrete is cast against earth.

Spacing of Bars.   Clear distance between bars shall not be less than 1.5 times the bar diameter, 1.33 times the maximum specified size for the coarse aggregate, or 1.5 in.

11-5. Design of Tied Columns

The 1963 ACI Code limits the axial compression load on a tied column to

$$P = 0.85[A_g(0.25f'_c + f_s p_g)]$$

in which P = maximum permissible axial load
A_g = gross area of the column
f'_c = ultimate compressive strength of the concrete
f_s = allowable compressive stress in the reinforcing,

taken as 40% of the yield stress but not to exceed
30,000 psi

p_g = percent of steel = A_g/A_s
A_s = cross-sectional area of the reinforcing

The following example illustrates the use of this formula for
the determination of the allowable load on a given column.

Example 1.   A 16-in. square tied column is reinforced with four
No. 10 bars; $f_c' = 4000$ psi and $f_s = 20,000$ psi. Find the safe load
for the column.

Solution: For use in the formula, we determine

$$A_g = 16 \times 16 = 256 \text{ in.}^2$$

$$p_g = \frac{A_s}{A_g} = \frac{(4 \times 1.27)}{256} = 0.0198$$

Then

$$P = 0.85[256(0.25 \times 4 + 20 \times 0.0198)] = 303.8 \text{ k}$$

In most building structures, concrete columns will sustain
some computed bending moment in addition to the axial compression load (see Fig. 11-3). Even when a computed moment is not
present, however, it is well to consider some amount of acciden-
tal eccentricity or other source of moment. It is recommended,
therefore, that the maximum safe load be limited to that given for
a minimum eccentricity of 10% of the column dimension.

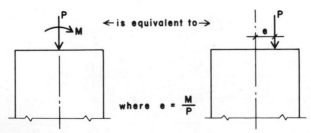

FIGURE 11-3.   Moment plus axial compression on a column visualized as an
equivalent eccentric load.

Figure 11-4 gives safe loads for a selected number of sizes of square tied columns. Loads are given for various degrees of eccentricity, which is a means for expressing axial load and bending moment combinations. The computed moment on the column is translated into an equivalent eccentric loading, as shown in Fig. 11-3. Data for the curves was computed by using 40% of the load

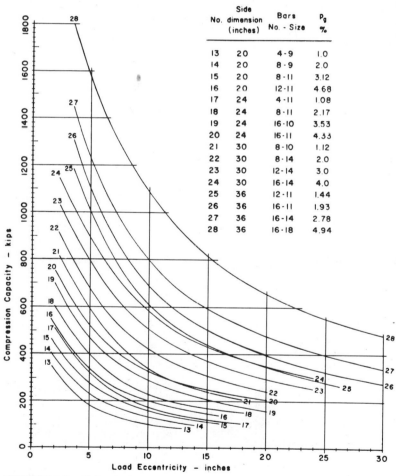

No.	Side dimension (inches)	Bars No. - Size	P_g %
13	20	4 - 9	1.0
14	20	8 - 9	2.0
15	20	8 - 11	3.12
16	20	12 - 11	4.68
17	24	4 - 11	1.08
18	24	8 - 11	2.17
19	24	16 - 10	3.53
20	24	16 - 11	4.33
21	30	8 - 10	1.12
22	30	8 - 14	2.0
23	30	12 - 14	3.0
24	30	16 - 14	4.0
25	36	12 - 11	1.44
26	36	16 - 11	1.93
27	36	16 - 14	2.78
28	36	16 - 18	4.94

FIGURE 11-4. Safe service loads for square tied columns with f'_c = 4 ksi and f_y = 60 ksi.

determined by strength design methods, as required by the 1977 ACI Code.

For the column in Example 1, it may be noted that Fig. 11-4 yields a maximum value of approximately 260 k, which is only about 80% of the value previously determined by the working stress formula. The discrepancy occurs because the code requires a minimum eccentricity for all columns; thus the curves in Fig. 11-4 do not begin at zero eccentricity. In addition, the requirement that only 40% of the load determined by strength methods

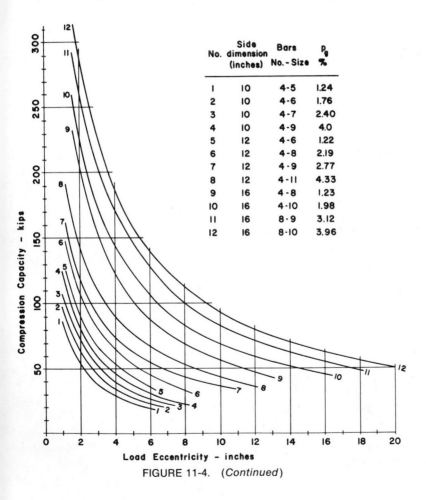

No.	Side dimension (inches)	Bars No.-Size	p_g %
1	10	4-5	1.24
2	10	4-6	1.76
3	10	4-7	2.40
4	10	4-9	4.0
5	12	4-6	1.22
6	12	4-8	2.19
7	12	4-9	2.77
8	12	4-11	4.33
9	16	4-8	1.23
10	16	4-10	1.98
11	16	8-9	3.12
12	16	8-10	3.96

FIGURE 11-4. (*Continued*)

be used places a rather high safety factor on the working stress method. To say the least, the code does not favor the working stress method in this case.

The following examples illustrate the use of Fig. 11-4 for the design of tied columns.

Example 2. A column with $f'_c = 4$ ksi and steel with $f_y = 60$ ksi sustains an axial compression load of 400 k. Find the minimum practical column size if reinforcing is a maximum of 4% and the maximum size if reinforcing is a minimum of 1%.

Solution: Using Fig. 11-4a, we find from the sizes given:

Minimum column is 20 in.^{2} with 8 No. 9 (Curve No. 14).

Maximum capacity is 410 kips, $p_g = 2.0\%$

Maximum size is 24 in.^{2} with 4 No. 11 (Curve No. 17).

Maximum capacity is 510 kips, $p_g = 1.08\%$.

It should be apparent that it is possible to use an 18-in. or 19-in. column as the minimum size and to use a 22-in. or 23-in. column as the maximum size. Since these sizes are not given in the figure, we cannot verify them for certain without using strength design procedures.

Example 3. A square tied column with $f'_c = 4$ ksi and steel with $f_y = 60$ ksi sustains an axial load of 400 k and a bending moment of 200 k-ft. Determine the minimum size column and its reinforcing.

Solution: We first determine the equivalent eccentricity, as shown in Fig. 11-3. Thus

$$e = \frac{M}{P} = \frac{200 \times 12}{400} = 6 \text{ in.}$$

Then, from Fig. 11-4, we find:

Minimum size is 24 in. square with 16 No. 10 bars.

Capacity at 6 in. eccentricity is 410 k.

Problems 11-5-A* to 11-5-E. Using Fig. 11-4, pick the minimum size square tied column and its reinforcing for the following combinations of axial load and bending moment.

	Axial compressive load (k)	Bending moment (k-ft)
A	100	25
B	100	50
C	150	75
D	200	100
E	300	150

11-6. Bar Layouts for Rectangular Tied Columns

Usually a number of possible combinations of reinforcing bars may be assembled to satisfy the steel area requirement for a given column. Aside from providing for the area, the number of bars must also work reasonably in the layout of the column. Figure 11-5 shows a number of tied columns with various number of bars. When a column is small, the preferred choice is usually that of the simple four-bar layout, with one bar in each corner and a single peripheral tie. As the column gets larger, the distance between the corner bars gets larger, and it is best to use more bars so that the reinforcing is spread out around the column periphery. For a symmetrical layout and the simplest of tie layouts, the best choice is for numbers that are multiples of four, as shown in

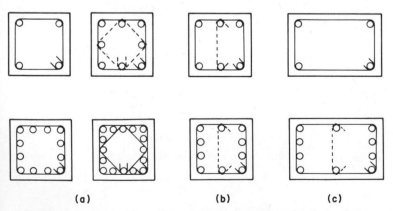

(a)   (b)   (c)

FIGURE 11-5. Typical bar placement and tie patterns for tied columns.

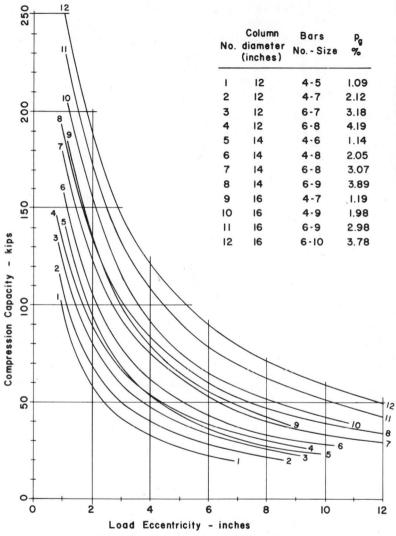

No.	Column diameter (inches)	Bars No.-Size	p_g %
1	12	4-5	1.09
2	12	4-7	2.12
3	12	6-7	3.18
4	12	6-8	4.19
5	14	4-6	1.14
6	14	4-8	2.05
7	14	6-8	3.07
8	14	6-9	3.89
9	16	4-7	1.19
10	16	4-9	1.98
11	16	6-9	2.98
12	16	6-10	3.78

FIGURE 11-6. Safe service loads for round tied columns with $f'_c = 4$ ksi and $f_y = 60$ ksi.

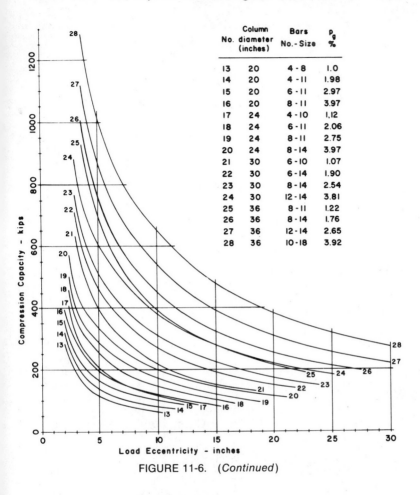

No.	Column diameter (inches)	Bars No.-Size	p_g %
13	20	4 - 8	1.0
14	20	4 - 11	1.98
15	20	6 - 11	2.97
16	20	8 - 11	3.97
17	24	4 - 10	1.12
18	24	6 - 11	2.06
19	24	8 - 11	2.75
20	24	8 - 14	3.97
21	30	6 - 10	1.07
22	30	6 - 14	1.90
23	30	8 - 14	2.54
24	30	12 - 14	3.81
25	36	8 - 11	1.22
26	36	8 - 14	1.76
27	36	12 - 14	2.65
28	36	10 - 18	3.92

FIGURE 11-6. (*Continued*)

Fig. 11-5*a*. The number of additional ties required for these layouts depends on the size of the column and the considerations discussed in Section 11-4.

An unsymmetrical bar arrangement is not necessarily bad, even though the column and its construction details are otherwise not oriented differently on the two axes. In situations where moments may be greater on one axis, the unsymmetrical layout is actually preferred; in fact, the column shape will also be more

effective if it is unsymmetrical, as shown for the oblong shapes in Fig. 11-5c.

11-7. Design of Round Columns

Round columns may be designed and built as spiral columns as described in Section 11-3, or they may be developed as tied columns with the bars placed in a circle and held by a series of round circumferential ties. Because of the cost of spirals, it is often more economical to use the tied columns, so they are often used unless the additional strength or other behavioral characteristics of the spiral column are required. In such cases, the column is usually designed as a square column using the square shape that can be included within the round form. It is thus possible to use a four-bar column for small-diameter, round column forms.

Figure 11-6 gives safe loads for round columns that are designed as tied columns. Load values have been adapted from values determined by strength design methods. The curves in Fig. 11-6 are similar to those for the square columns in Fig. 11-4, and their use is similar to that demonstrated in Examples 2 and 3 of Section 11-5.

Problems 11-7-A to 11-7-E.   Using Fig. 11-6, pick the minimum size round column and its reinforcing for the load and moment combinations in Problem 11-5.

11-8. Slenderness Effects in Compression Members

Most concrete columns in building construction are relatively stout. Although the code provides for reduction of axial compression on the basis of slenderness, the reductions do not become significant until the ratio of the column height to its least lateral dimension exceeds about 12. For slenderness beyond this ratio, the code reductions should be considered.

12

Footings

||||||||||||||||||||||||||||||||||||||||||||||||

12-1. Building Foundations

The foundations of a building are those structural elements that support the superstructure and are placed below grade. Foundations consist of wall and column footings of various types, support directly on the foundation bed or on piles. The primary purpose of a footing is to spread the loads so that the allowable bearing capacity of the foundation bed is not exceeded. With the exception of bedrock, a certain amount of settlement is present for all foundation beds. It is important that the settlement be as little as possible and that, if there is settlement, it is uniform throughout the entire structure. Another essential requirement for foundations is that the lines of action of the loads, whenever possible, coincide with centers of the foundations; that is to say, the pressure on the foundation bed should be uniformly distributed. If this condition does not exist, unequal pressures result and there is a tendency toward unequal settlement.

12-2. Bearing Pressure for Design

Before determining the required dimensions of a foundation, the designer must ascertain the allowable bearing capacity of the

foundation bed. In cities in which experience and tests have established the allowable strengths of various foundation soils local building codes may be consulted to determine the bearing capacities to be used in design. In the absence of such information, or for conditions in which the nature of the soil is unknown, borings or load tests should be made. For sizable structures borings at the site should always be made.

Most building codes allow for the use of so-called presumptive bearing pressures for design. These are average values—usually on the conservative side—that may be used for properly identified soils classified by groupings given by the codes. Table 12-1 gives such average values as adapted from the 1982 edition of the *Uniform Building Code* (Ref. 8).

For a more complete discussion of soil behaviors, of determination of soil properties, and of the design of ordinary foundation elements, the reader is referred to *Simplified Design of Building Foundations* (Ref. 6).

TABLE 12-1.   Allowable Bearing Capacities of Various Foundation Beds (psf)[a]

	Allowable pressure	
Description of materials	psf	kN/m²
Massive crystalline bedrock	4000	192
Sedimentary and foliated rock	2000	96
Sandy gravel and/or gravel	2000	96
Sand, silty sand, clayey sand, silty gravel, and clayey gravel	1500	72
Clay, sandy clay, silty clay, and clayey silt	1000	48

Source: Uniform Building Code, 1982 ed. (Ref. 8), with permission of the publishers, International Conference of Building Officials.

[a] Values are for footings with minimum width of 12 in. [0.3m] and minimum depth into natural grade of 12 in. Except for last category, an increase of 20% is allowed for each additional foot of width and/or depth to a maximum of three times the designated value.

12-3. Independent Column Footings

The great majority of independent or isolated column footings are square in plan, with reinforcing consisting of two sets of bars at right angles to each other. This is known as two-way reinforcement. The column may be placed directly on the footing block, or it may be supported by a pedestal. A pedestal, or pier, is a short, wide compression block that serves to reduce the punching effect on the footing. For steel columns a pier may also serve to raise the bottom of the steel column above ground level.

The design of a column footing is usually based on the following considerations:

1. Maximum Soil Pressure. The sum of the superimposed load on the footing and the weight of the footing must not exceed the limit for bearing pressure on the supporting material. The required total plan area of the footing is determined on this basis.

2. Control of Settlement. Where buildings rest on highly compressible soil, it may be necessary to select footing areas that assure a uniform settlement of all the building columns rather than to strive for a maximum use of the allowable soil pressure.

3. Size of the Column. The larger the column, the less will be the shear, flexural, and bond stresses in the footing, since these are developed by the cantilever effect of the footing projection beyond the edges of the column.

4. Shear Stress Limit for the Concrete. For square-plan footings this is usually the only critical stress condition for the concrete. In order to reduce the required amount of reinforcing, the footing depth is usually established well above that required by the flexural stress limit for the concrete.

5. Flexural Stress and Development Length Limits for the Bars. These are considered on the basis of the moment developed in the cantilevered footing at the face of the column.

6. Footing Thickness for Development of Column Reinforcing. When a footing supports a reinforced concrete

column, the compressing force in the column bars must be transferred to the footing by bond stress—called doweling of the bars. The thickness of the footing must be sufficient for the necessary development length of the column bars.

12-4. Design of a Column Footing

The following example illustrates the design process for a simple, square column footing.

Example.   A 16-in. [406-mm] square concrete column exerts a load of 240 [1068 kN] on a square column footing. Determine the footing dimensions and the necessary reinforcing using the following data: f'_c = 3 ksi [20.7 MPa], Grade 40 bars with f_y = 40 ksi [276 MPa] and f_s = 20 ksi [138 MPa], maximum permissible soil pressure = 4000 psf [192 kN/m^2].
Solution: The first decision to be made is that of the height, or thickness, of the footing. This has to be a raw first guess unless the dimensions of similar footings are known. In practice this knowledge is generally available from previous design work or from handbook tables. In lieu of this, a reasonable guess is made, the design work is performed, and an adjustment is made if the assumed thickness proves inadequate. We will assume a footing thickness of 20 in. [508 mm] for a first try for this example.

The footing thickness establishes the weight of the footing on a per-square-foot basis. This weight is then subtracted from the maximum permissible soil pressure, and the net value is then usable for the superimposed load on the footing. Thus

$$\text{footing weight} = \frac{20}{12}\,(150\text{ psf}) = 250\text{ psf }[12\text{ kN/m}^2]$$

net usable pressure = 4000 − 250 = 3750 psf [180 kN/m^2]

$$\text{required footing plan area} = \frac{240{,}000}{3750} = 64\text{ ft}^2 \left[\frac{1068}{180} = 5.93\text{ m}^2\right]$$

and the length of the side of the square footing,

$$L = \sqrt{64} = 8\text{ft }[\sqrt{5.93} = 2.44\text{ m}]$$

Two shear stress situations must be considered for the concrete. The first occurs as ordinary beam shear in the cantilevered portion and is computed at a critical section at a distance d (effective depth of the beam) from the face of the column as shown in Fig. 12-1a. The shear stress at this section is computed in the same manner as for a beam, as discussed in Section 7-3, and the stress limit is $v_c = \sqrt{f_c'}$. The second shear stress condition is that of peripheral shear, or so-called "punching" shear, and is investigated at a circumferential section around the column at a distance of $d/2$ from the column face as shown in Fig. 12-1b. For this condition the allowable stress is $v_c = 2.0\sqrt{f_c'}$.

With two-way reinforcing, it is necessary to place the bars in one direction on top of the bars in the other direction. Thus, although the footing is supposed to be the same in both directions, there are actually two different d distances—one for each layer of

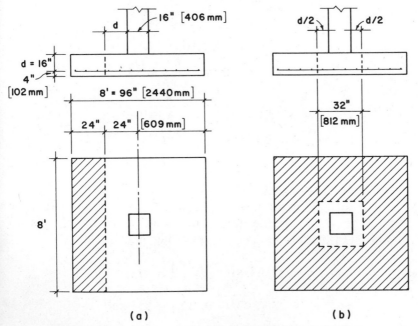

(a)   (b)

FIGURE 12-1.   Shear considerations for the column footing.

bars. It is common practice to use the average of these two distances for the design value of d; that is, d = the footing thickness less the sum of the concrete cover and the bar diameter. With the bar diameter as yet undetermined, we will assume an approximate d of the footing thickness less 4 in. [102 mm] (a concrete cover of 3 in. plus a No. 8 bar). For the example this becomes

$$d = t - 4 = 20 - 4 = 16 \text{ in. [406 mm]}$$

It should be noted that it is the *net* soil pressure that causes stresses in the footing, since there will be no bending or shear in the footing when it rests alone on the soil. We thus use the net soil pressure of 3750 psf [180 kN/m²] to determine the shear and bending effects for the footing.

For the beam shear investigation, we determine the shear force generated by the net soil pressure acting on the shaded portion of the footing plan area shown in Fig. 12-1a. Thus

$$V = 3750 \times 8 \times \frac{24}{12}$$

$$= 60,000 \text{ lb } [180 \times 2.44 \times 0.609 = 267.5 \text{ kN}]$$

and, using the formula for shear stress in a beam (Section 7-3),

$$v = \frac{V}{bd} = \frac{60,000}{96 \times 16} = 39.1 \text{ psi } \left[\frac{0.2675}{2.44 \times 0.406} = 0.270 \text{ MPa} \right]$$

which is compared to the allowable stress of

$$v_c = 1.1\sqrt{f'_c} = 1.1\sqrt{3000} = 60 \text{ psi [0.414 MPa]}$$

indicating that this condition is not critical.

For the peripheral shear investigation, we determine the shear force generated by the net soil pressure acting on the shaded portion of the footing area shown in Fig. 12-1b. Thus

$$V = 3750 \left[(8)^2 - \left(\frac{32}{12}\right)^2 \right] = 213,333 \text{ lb}$$

$$[V = 180(2.44^2 - 0.812^2) = 953 \text{ kN}]$$

Shear stress for this case is determined with the same formula

as for beam shear, with the dimension b being the total peripheral circumference. Thus

$$v = \frac{V}{bd} = \frac{213{,}333}{(4 \times 32) \times 16} = 104.2 \text{ psi}$$

$$\left[v = \frac{0.953}{(4 \times 0.812 \times 0.406} = 0.723 \text{ MPa} \right]$$

which is compared to the allowable stress of

$$v_c = 2\sqrt{f'_c} = 2\sqrt{3000} = 109.5 \text{ psi } [0.755 \text{ MPa}]$$

This computation indicates that the peripheral shear stress is not critical, but since the actual stress is quite close to the limit, the assume thickness of 20 in. is probably the least full-inch value that can be used. Flexural stress in the concrete should also be considered, although it is seldom critical for a square footing. One way to verify this is to compute the balanced moment capacity of the section with $b = 96$ in. and $d = 16$ in. Using the factor for a balanced section from Table 6-2, we find

$$M_R = Rbd^2 = 0.226(96)(16)^2 = 5554 \text{ k-in or } 463 \text{ k-ft}$$

which may be compared with the actual moment computed in the next step.

For the reinforcing we consider the stresses developed at a section at the edge of the column as shown in Fig. 12-2. The cantilever moment for the 40-in. [1016-mm] projection of the footing beyond the column is

$$M = 3750 \times 8 \times \frac{40}{12} \times \frac{1}{2} \left(\frac{40}{12} \right) = 166{,}667 \text{ lb-ft}$$

$$\left[M = 180 \times 2.44 \times 1.016 \times \frac{1.016}{2} = 227 \text{ kN/m} \right]$$

Using the formula for required steel area in a beam, with a conservative guess of 0.9 for j, we find (see Section 6-5)

$$A_s = \frac{M}{f_s jd} = \frac{166{,}667 \times 12}{20 \times 0.9 \times 16 \times 10^3} = 6.95 \text{ in.}^2 \text{ [4502 mm}^2\text{]}$$

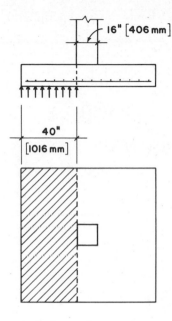

FIGURE 12-2. Bending and development length considerations for the column footing.

This requirement may be met by various combinations of bars, such as those in Table 12-2. Data for consideration of the development length and the center-to-center bar spacing is also given in the table. The flexural stress in the bars must be developed by the embedment length equal to the projection of the bars beyond the column edge, as discussed in Section 8-3. With a minimum of 2 in. [51 mm] of concrete cover at the edge of the footing, this length is 38 in. [965 mm]. The required development lengths indicated in the table are taken from Table 8-1, it may be noted that all of the combinations in the table are adequate in this regard.

If the distance from the edge of the footing to the first bar at each side is approximately 3 in. [76 mm], the center-to-center distance for the two outside bars will be 96 − 2(3) = 90 in. [2286 mm], and with the rest of the bars evenly spaced, the spacing will be 90 divided by the number of total bars less one. This value is shown in the table for each set of bars. The maximum permitted spacing is 18 in. [457 mm], and the minimum should be a distance

that is adequate to permit good flow of the wet concrete between the two-way grid of bars—say 4 in. [102 mm] or more.

All of the bar combinations in Table 12-2 are adequate for the footing. Many designers prefer to use the largest possible bar, as this reduces the number of bars that must be handled and supported during construction. On this basis, the footing will be the following:

8 ft² by 20 in. thick with six No. 10 bars each way.

Problem 12-3-A. Design a square footing for a 14-in. [356-mm] square concrete column with a load of 219 k [974 kN]. The maximum permissible soil pressure is 3000 psf [144 kN/m²]. Use concrete with $f'_c = 3$ ksi [20.7 MPa] and reinforcing of Grade 40 bars with $f_y = 40$ ksi [276 MPa] and $f_s = 20$ ksi [138 MPa].

12-5. Load Tables for Column Footings

For ordinary situations we often design square column footings by using data from tables in various references. Even when special circumstances make it necessary to perform the type of design illustrated in Section 12-4, such tables will assist in making a first guess for the footing dimensions.

Table 12-3 gives the allowable superimposed load for a range of footings and soil pressures. This material has been adapted from a more extensive table in *Simplified Design of Building Foundations* (Ref. 6). Designs are given for concrete strengths of 2000

TABLE 12-2. Reinforcing Alternatives for the Column Footing

Number and size of bars	Area of steel provided		Required development length[a]		Center-to-center spacing	
	in.²	mm²	in.	mm	in.	mm
12 No.7	7.20	4645	18	457	8.2	208
9 No. 8	7.11	4687	23	584	11.3	286
7 No. 9	7.00	4516	29	737	15	381
6 No. 10	7.62	4916	37	940	18	458

[a] From Table 8-1; values for "other bars," $f_y = 40$ ksi, $f'_c = 3$ ksi.

TABLE 12-3. Allowable Loads on Square Cclumn Footings (see Fig. 12-3)

Maximum soil pressure (lb/ft²)	Minimum column width t (in.)	$f'_c = 2000$ psi				$f'_c = 3000$ psi			
		Allowable load on footing[a] (k)	Footing dimensions h (in.)	Footing dimensions w (ft)	Reinforcing each way	Allowable load on footing (k)	Footing dimensions h (in.)	Footing dimensions w (ft)	Reinforcing each way
1000	8	7.9	10	3.0	2 No. 3	7.9	10	3.0	2 No. 3
	8	10.7	10	3.5	3 No. 3	10.7	10	3.5	3 No. 3
	8	14.0	10	4.0	3 No. 4	14.0	10	4.0	3 No. 4
	8	17.7	10	4.5	4 No. 4	17.7	10	4.5	4 No. 4
	8	22	10	5.0	4 No. 5	22	10	5.0	4 No. 5
	8	31	10	6.0	5 No. 6	31	10	6.0	5 No. 6
	8	42	12	7.0	6 No. 6	42	11	7.0	7 No. 6

1500	8	12.4	10	3.0	3 No. 3	12.4	10	3.0	3 No. 3
	8	16.8	10	3.5	3 No. 4	16.8	10	3.5	3 No. 4
	8	22	10	4.0	4 No. 4	22	10	4.0	4 No. 4
	8	28	10	4.5	4 No. 5	28	10	4.5	4 No. 5
	8	34	12	5.0	5 No. 5	34	11	5.0	6 No. 5
	8	48	14	6.0	6 No. 6	49	13	6.0	6 No. 6
	8	65	16	7.0	7 No. 6	65	15	7.0	6 No. 7
	8	83	18	8.0	7 No. 7	84	16	8.0	7 No. 7
		103		9.0	8 No. 7	105		9.0	10 No. 7
2000	8	17	10	3.0	4 No. 3	17	10	3.0	4 No. 3
	8	23	10	3.5	4 No. 4	23	10	3.5	4 No. 4
	8	30	10	4.0	6 No. 4	30	10	4.0	6 No. 4
	8	37	11	4.5	5 No. 5	38	11	4.5	6 No. 5
	8	46	12	5.0	6 No. 5	46	13	5.0	5 No. 6
	8	65	14	6.0	6 No. 6	66	15	6.0	7 No. 6
	8	88	16	7.0	8 No. 6	89	17	7.0	7 No. 7
	8	113	18	8.0	8 No. 7	114	19	8.0	9 No. 7
	8	142	20	9.0	8 No. 8	143	20	9.0	8 No. 8
	10	174	21	10.0	9 No. 8	175		10.0	10 No. 8

TABLE 12-3. (Continued)

Maximum soil pressure (lb/ft²)	Minimum column width t (in.)	f'c = 2000 psi				f'c = 3000 psi			
		Allowable load on footing^a (k)	Footing dimensions h (in.)	w (ft)	Reinforcing each way	Allowable load on footing (k)	Footing dimensions h (in.)	w (ft)	Reinforcing each way
3000	8	26	10	3.0	3 No. 4	26	10	3.0	3 No. 4
	8	35	10	3.5	4 No. 5	35	10	3.5	4 No. 5
	8	45	12	4.0	4 No. 5	46	11	4.0	5 No. 5
	8	57	13	4.5	6 No. 5	57	12	4.5	6 No. 5
	8	70	14	5.0	5 No. 6	71	13	5.0	6 No. 6
	8	100	17	6.0	7 No. 6	101	15	6.0	8 No. 6
	10	135	19	7.0	7 No. 7	136	18	7.0	8 No. 7
	10	175	21	8.0	10 No. 7	177	19	8.0	8 No. 8
	12	219	23	9.0	9 No. 8	221	21	9.0	10 No. 8
	12	269	25	10.0	11 No. 8	271	23	10.0	10 No. 9
	12	320	28	11.0	11 No. 9	323	26	11.0	12 No. 9
	14	378	30	12.0	12 No. 9	381	28	12.0	11 No. 10

f_c'	t (in)	Load	Width (ft)	n	Reinforcement	Load	n	Width (ft)	Reinforcement
4000	8	35	3.0	10	4 No. 4	35	10	3.0	4 No. 4
	8	47	3.5	12	4 No. 5	47	11	3.5	4 No. 5
	8	61	4.0	13	5 No. 5	61	12	4.0	6 No. 5
	8	77	4.5	15	5 No. 6	77	13	4.5	6 No. 6
	8	95	5.0	16	6 No. 6	95	15	5.0	6 No. 6
	8	135	6.0	19	8 No. 6	136	18	6.0	7 No. 7
	10	182	7.0	22	8 No. 7	184	20	7.0	9 No. 7
	10	237	8.0	24	9 No. 8	238	22	8.0	9 No. 8
	12	297	9.0	26	10 No. 8	299	24	9.0	9 No. 9
	12	364	10.0	29	13 No. 8	366	27	10.0	11 No. 9
	14	435	11.0	32	12 No. 9	440	29	11.0	11 No. 10
	14	515	12.0	34	14 No. 9	520	31	12.0	13 No. 10
	16	600	13.0	36	17 No. 9	606	33	13.0	15 No. 10
	16	688	14.0	39	15 No. 10	696	36	14.0	14 No. 11
	18	784	15.0	41	17 No. 10	793	38	15.0	16 No. 11

[a] *Note*: Allowable loads do not include the weight of the footing, which has been deducted from the total bearing capacity. Criteria: $f_s = 20$ ksi, $v_c = 1.1\sqrt{f_c'}$ for beam shear, $v_c = 2\sqrt{f_c'}$ for peripheral shear.

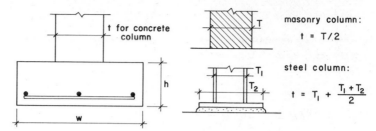

FIGURE 12-3.   Reference figure for Table 12-3.

and 3000 psi. The low strength of 2000 psi is sometimes used for small buildings, since many building codes permit the omission of testing of the concrete if this value is used for design.

12-6.   Pedestals

A pedestal (also called a pier) is defined by the ACI Code as a short compression member whose height does not exceed three times its width. Pedestals are frequently used as transitional elements between columns and the bearing footings that support them. Figure 12-4 shows the use of pedestals with both steel and reinforced concrete columns. The most common reasons for use of pedestals are:

1.   To spread the load on top of the footing. This may relieve the intensity of direct bearing pressure on the footing or

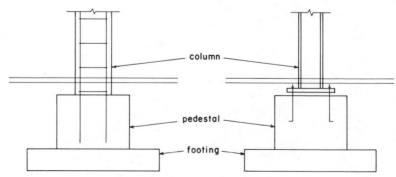

FIGURE 12-4.   Use of concrete pedestals.

may simply permit a thinner footing with less reinforcing due to the wider column.

2. To permit the column to terminate at a higher elevation where footings must be placed at depths considerably below the lowest parts of the building. This is generally most significant for steel columns.

3. To provide for the required development length of reinforcing in reinforced concrete columns, where footing thickness is not adequate for development within the footing.

Figure 12-5 illustrates the third situation described. Referring to Table 8-3, we may observe that a considerable development length is required for large diameter bars made from high grades of steel. If the minimum required footing does not have a thickness that permits this development, a pedestal may offer a reasonable solution. However, there are many other considerations to be made in the decision, and the column reinforcing problem is not the only factor in this situation.

If a pedestal is quite short with respect to its width (see Fig. 12-6a), it may function essentially the same as a column footing, with significant values for shear and bending stresses. This condition is likely to occur if the pedestal width exceeds twice the

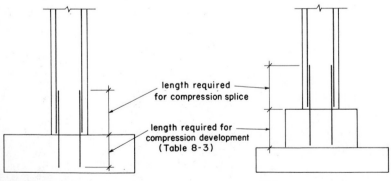

length required
for compression splice

length required for
compression development
(Table 8-3)

(a) Footing without Pedestal   (b) Footing with Pedestal

FIGURE 12-5.   Development length considerations for concrete columns: (a) without pedestals and (b) with pedestals.

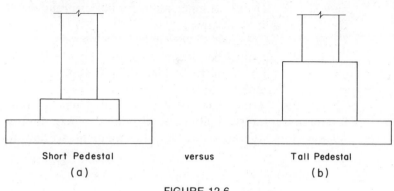

Short Pedestal   versus   Tall Pedestal
(a)   (b)

FIGURE 12-6.

column width and the pedestal height is less than one-half of the pedestal width. In such cases, the pedestal must be designed by the same procedures used for an ordinary column footing.

The following example illustrates the procedure for the design of a pedestal for a reinforced concrete column.

Example.   A 16-in. square tied column with f'_c of 4 ksi is reinforced with No. 10 bars of Grade 60 steel (F_y = 60 ksi). The column axial load is 200 k, and the allowable maximum soil pressure is 4000 psf. Design a footing and a pedestal, using f'_c = 3 ksi and Grade 40 reinforcing with f_y = 40 ksi.

Solution: For an approximate idea of the required footing, we may refer to Table 12-3 and observe the following.

> 8-ft square footing, 22 in. thick, nine No. 8 each way.
> Allowable load on footing: 238 kips.
> Designed for column width of 10 in.

From Table 8-3, for No. 10 bar, Grade 60, we observe the following:

$$f'_c = 3 \text{ ksi}, \qquad l_d = 27.8 \text{ in.}$$

From these observations, we may conclude that:

1.   The minimum required footing for the 16-in. column with

200-k load will be slightly smaller than that taken from the table. Thus it will not be adequate for development of the column bars.

2. If a pedestal is used, it must be at least 28 in. high to develop the column bars.
3. With a pedestal slightly wider than the column, the footing thickness may be additionally reduced, if shear stress is the critical design factor for the footing thickness.

One option in this case is to simply forget about a pedestal and increase the footing thickness to that required for development of the column bars. This means an increase from around 20 in. up to 31 in., giving the necessary 28 in. of development plus 3 in. of cover. Let us therefore consider the possibility of a footing that is 7.5 ft square and 31 in. thick. Then,

$$\text{design soil pressure} = \frac{200,000}{(7.5)^2} = 3556 \text{ psf}$$

Adding the weight of the footing to this, the total soil pressure becomes

$$3556 + \frac{31}{12}(150) = 3944 \text{ psf}$$

which is less than the allowable of 4000 psf, so the footing width is adequate.

Shear stress is obviously not a critical concern, so we proceed to determine the required reinforcing. Figure 12-7 indicates the basis for determining the cantilever moment, and we thus compute the following.

$$M = 3.556 \times 7.5 \times \left(\frac{37}{12}\right)^2 \times \frac{1}{2} = 127 \text{ k-ft}$$

$$A_s = \frac{M}{f_s j d} = \frac{127 \times 12}{20(0.9)(27)}$$

$$= 3.14 \text{ in.}^2$$

Try six No. 7 bars:

$$A_s = 3.6 \text{ in.}^2$$

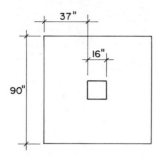

FIGURE 12-7.

Table 8-1 will indicate that the 37-in. projection is adequate for development of the No. 7 bars. It may be noted that this is considerably less reinforcing than that given for the footing taken from Table 12-3.

If it is desired to use a pedestal, we consider the use of the one shown in Fig. 12-8. The 28-in. height shown is the minimum established previously for the development of the column bars. The height could be increased to as much as 96 in. (three times the width), if it is desired for other reasons. One such reason may be the presence of a better soil for bearing at a lower elevation.

A potential concern is that for the direct bearing of the column on the pedestal. If the pedestal is designed as an unreinforced member, the ACI Code permits a maximum bearing stress of

$$f_p = 0.3 f_c' \times \sqrt{\frac{A_2}{A_1}}$$

where A_1 is the actual bearing area (in our case the 16-in. square column area) and A_2 is the area of the pedestal cross section. The maximum usable value for $\sqrt{A_2/A_1}$ is 2.

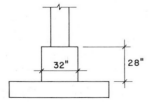

FIGURE 12-8.

In our example it will be found that the allowable stress thus determined is more than twice the value of the direct bearing, even if the latter is computed ignoring the portion of the load transferred by development of the column reinforcing. The only times that this condition is likely to be critical is when a pedestal with very low f'_c supports a column with very high f'_c and the pedestal width is only slightly greater than the column width. When the pedestal supports a steel column, however, this condition may be the basis for establishing the width of the pedestal.

Another consideration for bearing stress is that of the pedestal on the footing. In this case, using the same criteria described previously, the maximum allowable bearing stress will be either $0.3f'_c$ for the pedestal or $0.6f'_c$ (with the maximum value of $\sqrt{A_2/A_1}$) for the footing. This is also not critical for our example.

If the pedestal height exceeds its width, we recommend the use of a minimum of column reinforcing of not less than $A_s = 0.005A_g$. This should be installed with at least four bars, one in each corner, and a set of loop ties, just as with an ordinary tied column. For our short pedestal, this is of questionable necessity.

With the wide pedestal, the footing thickness can be reduced considerably, if the minimum thickness for shear stress is desired. The basis for consideration of peripheral shear stress is shown in Fig. 12-9a, and the computations are as follows. For the example, we have assumed a footing thickness of 14 in. with an effective depth of 10 in.

$$\text{weight of pedestal} = \frac{32 \times 32 \times 28}{1728}(150) = 2489 \text{ lb}$$

$$\text{design soil pressure} = \frac{202,489}{(7.5)^2} = 3600 \text{ psf}$$

$$V = 3.6\left\{(7.5)^2 - \left(\frac{42}{12}\right)^2\right\} = 158.4 \text{ k}$$

$$v = \frac{V}{bd} = \frac{158,400}{4 \times 42 \times 10} = 94.3 \text{ psi} < 110 \text{ psi}$$

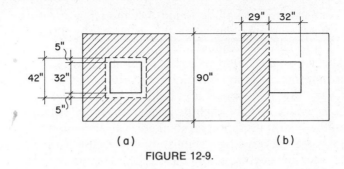

(a)   (b)

FIGURE 12-9.

The basis for consideration of the reinforcing is shown in Fig. 12-9*b*, and the computations are as follows.

$$M = 3.6 \times 7.5 \times \left(\frac{29}{12}\right)^2 \times \frac{1}{2} = 78.8 \text{ k-ft}$$

$$A_s = \frac{M}{f_s jd} = \frac{78.8 \times 12}{20 \ (0.9)(10)} = 5.25 \text{ in.}^2$$

This could be supplied by using seven No. 8 bars each way. It may be noted that this is considerably more reinforcing than that required for the thickened footing without the pedestal. Cost savings affected by the pedestal may thus be questionable, and its use may depend on its need for other purposes.

Problem 12-6-A. An 18-in. square tied column with $f'_c = 4$ ksi [27.6 MPa] is reinforced with No. 11 bars of Grade 60 steel with $f_y = 60$ ksi [414 MPa]. The column axial load is 260 k [1156 kN] and the allowable soil pressure is 3000 psf [134 kN/m²]. Using $f'_c = 3$ ksi [20.7 MPa] and Grade 40 bars with $f_y = 40$ ksi [276 MPa] and $f_s = 20$ ksi [138 MPa], design the following: (1) a footing without a pedestal and (2) a footing with a pedestal.

12-7. Wall Footings

Wall footings consist of concrete strips placed under walls. The most common type of wall footing is that shown in Fig. 12-10, consisting of a strip with a rectangular cross section placed in a symmetrical position with respect to the wall and projecting an equal distance as a cantilever from both faces of the wall. For soil

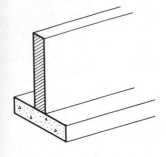

FIGURE 12-10. Continuous wall footing.

stress the critical dimension of the footing is the width of the footing bottom measured perpendicular to the wall face.

In most situations, the wall footing is utilized as a platform upon which the wall is constructed. Thus a minimum width for the footing is established by the wall thickness, the footing usually being made somewhat wider than the wall. With a concrete wall this additional width is used to support the wall forms while the concrete is poured. For masonry walls this added width assures an adequate base for the mortar bed for the first course of the masonry units. The exact additional width required for these purposes is a matter of judgment. For support of concrete forms, it is usually desirable to have at least a 3-in. projection; for masonry, the usual minimum is 2 in.

With relatively lightly loaded walls, the minimum width required for platform considerations may be more than adequate in terms of the allowable bearing stress on the soil. If this is the case, the short projection of the footing from the wall face will produce relatively insignificant transverse bending and shear stresses, permitting a minimal thickness for the footing and the omission of transverse reinforcing. Most designers prefer, however, to provide some continuous reinforcing in the long direction of the footing, even when none is used in the transverse direction. The purpose is to reduce shrinkage cracking and also to give some enhanced beamlike capabilities for spanning over soft spots in the supporting soil.

As the wall load increases, the increased width of the footing required to control soil stress eventually produces significant

transverse bending and shear in the footing. At some point this determines the required thickness for the footing and for required reinforcing in the transverse direction. If the footing is not reinforced in the transverse direction, the contolling stress is usually the transverse tensile bending stress in the concrete. If the footing has transverse reinforcing, the controlling concrete stress is usually the shear stress.

12-8. Design of a Wall Footing

The following example illustrates the procedure for design of a wall footing with transverse reinforcing.

Example.   Using concrete with f'_c = 2 ksi and Grade 40 reinforcing, design a wall footing for the following data: wall thickness = 6 in.; load on footing = 8750 lb/ft of wall length; maximum allowable soil pressure = 2000 psf.

Solution: For the wall footing, the only concrete stress of concern is that of shear. Compression stress in flexure is seldom critical due to design for shear and the desire for minimal transverse reinforcing. The usual design procedure consists of making a guess for the footing thickness and determining conditions to verify the guess. Code restrictions establish a minimum thickness of 10 in. Try

"ACF"
6" ꞩ7-86

$$h = 12 \text{ in.}$$

Then

$$\text{footing weight is 150 psf}$$

$$\text{usable soil pressure is } 2000 - 150 = 1850 \text{ psf}$$

$$\text{required width} = \frac{8750}{1850} = 4.73 \text{ ft or } 56.8 \text{ in.}$$

Try

$$\text{width} = 57 \text{ in. or } 4 \text{ ft} - 9 \text{ in.}$$

Then

$$\text{design soil pressure} = \frac{8750}{4.75} = 1842 \text{ psf}$$

With 3-in. cover and a No. 6 bar (a guess), the effective depth will be 8.625 in., say 8.6 in. approximately.

The ACI Code requires that shear be investigated as a beam sheer condition, with the critical section at a distance d (effective depth) from the face of the wall. This condition is shown in Fig. 12-11a. This is reasonably valid when the cantilever distance is larger than the footing thickness by a significant amount, but is questionable for short cantilevers. In fact, the code recommends that this shortened span not be used for brackets and short cantilevers. Therefore, we recommend that the critical section for shear be taken at the face of the wall unless the cantilever exceeds three times the overall footing thickness. However, if the latter assumption is made (short cantilever analysis), it is reasonable to use the full thickness of the footing for the stress computation rather than the effective depth of the cross section. Both cases are shown in Fig. 12-11, and we will show the computations for both.

Case 1. Shear at the d distance from the wall (see Fig. 12-11a).

Shear force:

$$V = (1842) \left(\frac{16.9}{12}\right) = 2594 \text{ lb}$$

Stress:

$$v_c = \frac{V}{bd} = \frac{2594}{(12)(8.6)} = 25 \text{ psi}$$

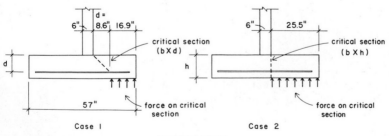

FIGURE 12-11.

Case 2. Shear at the wall face (see Fig. 12-11*b*).

$$V = (1842) \left(\frac{25.5}{12}\right) = 3914 \text{ lb}$$

Stress:

$$v_c = \frac{V}{bh} = \frac{3914}{(12)(12)} = 27 \text{ psi}$$

Both of these are well below the allowable stress of 1.1 $\sqrt{f'_c}$, which is the same as in the previous example: 49 psi. It is possible, therefore, to reduce the footing thickness if the shear stress is considered to be an important criterion. However, as has been discussed previously, reduction of cost in construction is usually obtained by minimizing the amount of reinforcing, and any reduction in the footing thickness will shorten the moment arm for the tension reinforcing, requiring an increase in steel area. It therefore becomes a matter of judgment about the ideal value for the footing thickness.

If we reduce the footing thickness to 11 in., a second try would proceed as follows:

$$\text{new footing weight} = \frac{11}{12}(150) = 137.5, \text{ say } 138 \text{ lb/ft}^2$$

$$\text{usable soil pressure } (p) = 2000 - 138 = 1862 \text{ lb/ft}^2$$

$$\text{required width } (w) = \frac{8750}{1862} = 4.70 \text{ ft or } 56.4 \text{ in.}$$

which does not change the footing width or design soil pressure.

$$\text{new } d = h - 3 - \frac{D}{2} = 11 - 3.375 = 7.625, \text{ say } 7.6 \text{ in.}$$

For the Case 2 shear stress, the shear force is the same as for the first try, and the new shear stress is

$$v_c = \frac{V}{bh} = \frac{3914}{(12)(11)} = 30 \text{ psi}$$

For the Case 1 shear stress, the shear section is now an inch closer to the wall and the shear force becomes

$$V = (1842) \left(\frac{17.9}{12}\right) = 2748 \text{ lb}$$

$$v_c = \frac{V}{bd} = \frac{2748}{(12)(7.6)} = 30 \text{ psi}$$

The bending moment to be used for concrete stress and determination of the steel area is

$$M = (3914) \left(\frac{25.5}{2}\right) = 49,903 \text{ lb-in.}$$

and the required steel area per foot of wall length is

$$A_s = \frac{M}{f_s j d} = \frac{49,903}{(20)(0.9)(7.6)} = 0.365 \text{ in.}^2$$

Since the steel area requirement has been determined in the same manner as for a slab, Table 9-4 may be used to select the bars and their spacing. The following should be considered in making the selection.

1. Maximum recommended spacing is 18 in.
2. Minimum recommended spacing is 6 in. to minimize the number of bars and allow for easy placing of the concrete during construction.
3. For proper development of the bars smaller bar sizes are usually preferable.

TABLE 12-4. Selection of Reinforcing for Example

Bar size	Area of bar (in.²)	Area required for flexure (in.²)	Spacing required (in.)	Selected spacing (in.)
3	0.11	0.365	3.6	3.5
4	0.20	0.365	6.6	6.5
5	0.31	0.365	10.2	10
6	0.44	0.365	14.5	14.5
7	0.60	0.365	19.7	19.5

TABLE 12-5. Allowable Loads on Wall Footings (see Fig. 12-12)

| Maximum soil pressure (lb/ft²) | Minimum wall thickness | | Allowable load on footing[a] (lb/ft) | Footing dimensions | | Reinforcing | |
	Concrete t (in.)	Masonry t (in.)		h (in.)	w (in.)	Long direction	Short direction
1000	4	8	2625	10	36	3 No. 4	No. 3 at 1€
	4	8	3062	10	42	2 No. 5	No. 3 at 1:
	6	12	3500	10	48	4 No. 4	No. 4 at 1€
	6	12	3938	10	54	3 No. 5	No. 4 at 1:
	6	12	4375	10	60	3 No. 5	No. 4 at 1(
	6	12	4812	10	66	5 No. 4	No. 5 at 1:
	6	12	5250	10	72	4 No. 5	No. 5 at 1▮
1500	4	8	4125	10	36	3 No. 4	No. 3 at 1(
	4	8	4812	10	42	2 No. 5	No. 4 at 1:
	6	12	5500	10	48	4 No. 4	No. 4 at 1▮
	6	12	6131	11	54	3 No. 5	No. 5 at 1:
	6	12	6812	11	60	5 No. 4	No. 5 at 1:
	6	12	7425	12	66	4 No. 5	No. 5 at 1▮
	8	16	8100	12	72	5 No. 5	No. 5 at 10▮
2000	4	8	5625	10	36	3 No. 4	No. 4 at 14▮
	6	12	6562	10	42	2 No. 5	No. 4 at 11
	6	12	7500	10	48	4 No. 4	No. 5 at 12▮
	6	12	8381	11	54	3 No. 5	No. 5 at 11
	6	12	9250	12	60	4 No. 5	No. 5 at 10▮
	8	16	10106	13	66	4 No. 5	No. 5 at 9
	8	16	10875	15	72	6 No. 5	No. 5 at 9
3000	6	12	8625	10	36	3 No. 4	No. 4 at 10▮
	6	12	10019	11	42	4 No. 4	No. 5 at 13
	6	12	11400	12	48	3 No. 5	No. 5 at 10
	6	12	12712	14	54	6 No. 4	No. 5 at 10
	8	16	14062	15	60	5 No. 5	No. 5 at 9
	8	16	15400	16	66	5 No. 5	No. 6 at 12
	8	16	16725	17	72	6 No. 5	No. 6 at 10

[a] *Note:* Allowable loads do not include the weight of the footing, which has been deducted from the total bearing capacity. Criteria: $f'_c = 2000$ psi, Grade 40 reinforcing, $v_c = 1.1 \sqrt{f'_c}$.

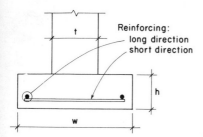

FIGURE 12-12.   Reference figure for
Table 12-5.

Table 12-4 presents a summary of the possible alternatives for
reinforcing in the transverse direction, as determined from the
data in Table 9-4. Our preference would be for the No. 5 bars at 10
in. center to center. Reference to Table 8-1 will show that development is more than adequate for these bars. (Note: Data is
not given in Table 8-1 for f'_c = 2000 psi; however, the table shows
only 12 in. required for f'_c = 3000 psi whereas over 23 in. is
available with our footing.)

Whether transverse reinforcing is provided or not, we recommend a minimum reinforcing for shrinkage stresses in the long
direction of the footing consisting of 0.0015 times the gross area
of the cross section. Thus:

$$A_s = 0.0015 \times 11 \times 57 = 0.94 \text{ in.}^2$$

This area can be supplied by using three No. 5 bars with a total
area of 0.93 in.^{2}

Problem 12-8-A.   Using concrete with f'_c = 2 ksi [13.8 MPa] and Grade 40 bars
with f_y = 40 ksi [276 MPa] and f_s = 20 ksi [138 MPa], design a wall footing for the
following data: wall thickness = 10 in. [254 mm]; load on footing = 12,000 lb/ft
[175 kN/m]; maximum soil pressure = 2000 psf [96 kN/m²].

12-9. Load Table for Wall Footings

Table 12-5 gives values for wall footings for four different soil
pressures. Table data was derived using the procedures illustrated in Section 12-8. Figure 12-12 shows the dimensions referred to in the table.

13

Walls

IIIIIIIIIIIIIIIIIIIIIIIIIIIIIIIIIIIIIIIIIIIIII

13-1. Introduction

Concrete walls are used for a variety of purposes in building construction. Classified with regard to their structural nature, the following types of walls are common.

1. Bearing Walls, Uniformly Loaded. These may be single story or multistory, carrying loads from floors, roofs, and/ or walls above.
2. Bearing Walls with Concentrated Loads. These are walls that provide support for beams or columns. In most cases they also support uniformly distributed loads.
3. Basement Walls, Earth Retaining. These are walls that occur at the exterior boundary between interior sublevel spaces and the surrounding earth. In addition to functioning as bearing walls (in most cases) they also span either vertically or horizontally as slabs to resist horizontal earth pressures.
4. Retaining Walls. This term is usually used to refer to walls that function to achieve grade-level changes, working essentially as vertical cantilevers to resist the horizontal earth pressures from the high side.
5. Shear Walls. These are walls that are used to brace the building against horizontal (lateral) forces due to wind or

earthquakes. The shear referred to is generated in the plane of the wall, as opposed to shear generated in slab-spanning action.

6. Freestanding Walls. These are walls used as fences or partitions, being supported only at their bases.
7. Grade Walls. These are walls that occur in buildings without basements; they function to support walls above grade and grade-level floor slabs. They may also function as grade beams or ties in buildings with isolated foundations consisting of column footings, piles, or piers.

It is possible, of course, for walls to serve more than one of these functions. Concrete walls are quite expensive when compared to other types of wall construction and, when used, are usually exploited for all their potential value for structural purposes.

In light building construction, concrete walls are sometimes built without reinforcing. The material in this chapter deals only with reinforced concrete walls.

13-2. General Requirements for Reinforced Concrete Walls

Regardless of their structural functions, a number of basic considerations apply to all walls. Some considerations of major concern are:

1. Wall Thickness. Nonstructural walls may be as thin as 4 in.; structural walls must be at least 6 in. thick. In general, slenderness ratio (unsupported height divided by thickness) should not exceed 25. A practical limit for a single pour (total height achieved in one continuous casting) is 15 times the wall thickness; taller walls will require multiple pours. Walls 10 in. or more in thickness should have two layers of reinforcing, one near each wall surface. Basement walls, foundation walls, and party walls must be at least 8 in. thick. Of course, the thickness must also be appropriate to the structural tasks.
2. Reinforcement. A minimum area of reinforcing equal to 0.0025 times the wall cross section must be provided in a

horizontal direction; 0.0015 in a vertical direction. A reduction is possible if bars No. 5 or smaller of Grade 60 or higher steel are used. As noted previously, two layers are required for walls 10 in. or more in thickness. The distribution of the total area required between the two layers depends on the wall functions.

3. Special Reinforcing Requirements.   General practice is to provide extra reinforcing at the top, bottom, ends, corners, intersections, and at openings in the wall. Suggested details for placement of reinforcing at these locations is given in the *ACI Detailing Manual* (Ref. 10) and various requirements are given in the ACI Code.

13-3. Bearing Walls

When the full wall cross section is utilized, bearing strength is limited as follows.

By working stress:

$$P = 0.30f'_c A_1$$

By strength methods:

$$P_u = 0.7(0.85f'_c A_1)$$

When the area developed in bearing is less than the total wall cross section, these loads may be increased by a factor equal to $\sqrt{A_2/A_1}$, but not more than 2. In this case A_1 is the actual bearing area and A_2 is the area of the full wall cross section.

When the resultant vertical compression force on a wall falls within the middle third of the wall thickness, the wall may be designed as an axial loaded column, using the following empirical formula with the strength method.

$$\phi P_{nw} = 0.55\phi f'_c A_g \left[1 - \left(\frac{l_c}{40h}\right)^2\right]$$

in which $\phi = 0.70$

P_{nw} = nominal axial load strength of wall
A_g = the effective area of the wall cross section
l_c = vertical distance between lateral supports
h = overall thickness of the wall

If the wall carries concentrated loads, the length of the wall to be considered as effective for each load shall not exceed the center-to-center distance between the loads nor the actual width of bearing plus four times the wall thickness.

The following example illustrates the procedure for design of a wall with concentrated loads. Design for a uniformly distributed load is essentially the same except that bearing stress and reduced effective area considerations need not be made.

Example. A reinforced concrete bearing wall supports a floor system consisting of precast single tees spaced 8 ft 0 in. on centers. The stem of each T-section is 8 in. wide, but the bearing width is taken as 7 in. to allow for beveled bottom edges. The tees will bear on the full thickness of the wall. The height of the wall is 11 ft 6 in., and the reaction of each tee due to service loads is 22 k for dead load and 12 k for live load. Design the wall in accordance with the following specification data: $f'_c = 4000$ psi and $f_y = 40,000$ psi. The reader should make a sketch to show the physical conditions described.

Solution: (1) The factored value of the reaction of one single T-section is

$$P_u = 1.4P_d + 1.7P_l$$

$$P_u = (1.4 \times 22) + (1.7 \times 12) = 51.2 \text{ k}$$

and

$$\frac{P_u}{\phi} = \frac{51.2}{0.70} = 73.1 \text{ k}$$

(2) Assume the minimum wall thickness of $h = 6$ in. and check the bearing stress f_b. Letting b' equal the bearing width of the tee stem,

$$f_b = 73,100 \div (7 \times 6) = 1740 \text{ psi}$$

allowable $f_b = 0.85\phi f'_c = 0.85 \times 0.70 \times 4000 = 2380$ psi

Because f_b is less than the allowable value, bearing on the wall is not critical.

(3) Determine the effective horizontal length of wall. This will

be controlled by the bearing width of the T-section plus four times the wall thickness or

$$b' + 4h = 7 + (4 \times 6) = 31 \text{ in.}$$

(4) Check the l_c/h ratio to see that it does not exceed 25.

$$\frac{l_c}{h} = \frac{11.5 \times 12}{6} = 23 < 25 \qquad \text{OK}$$

Therefore the minimum thickness of $h = 6$ in. is tentatively adopted.

(5) With the full bearing of the T-sections on the wall (and assuming that their deflection will not be sufficient to move the center of reactions outside the middle third of the wall) the "reasonably concentric" loading condition may be considered satisfied.

(6) Determine the allowable capacity of the wall from Code Eq. (14-1). This equation may be written

$$\frac{P_u}{\phi} = 0.55 f'_c A_g \left[1 - \left(\frac{l_c}{40h} \right)^2 \right]$$

Expressing f'_c in kips per square inch, l_c and h in feet, and noting that $A_g = 6 \times 31 = 186$ in.^{2},

$$\frac{P_u}{\phi} = 0.55 \times 4 \times 186 \left[1 - \left(\frac{11.5}{40} \times \frac{2.0}{0.5} \right)^2 \right]$$

$$\frac{P_u}{\phi} = 409 \times [1 - (0.575)^2] = 274 \text{ k}$$

which is greater than the required capacity of $P_u/\phi = 73.1$ k determined in Step 1. Therefore the capacity of the 6-in.-thick wall is adequate and provides a suitable margin for possible effect of eccentricity.

(7) Select the reinforcement. Because the required amounts of both vertical and horizontal steel are expressed as $\rho \times A_g$, we may work with one linear foot of wall instead of using the effec-

tive horizontal length of 31 in.; for this purpose, A_g then becomes $6 \times 12 = 72$ in.^{2}:

$$\text{vertical } A_s = 0.0015 \times 72 = 0.108 \text{ in.}^2 \text{ per linear foot}$$

$$\text{horizontal } A_s = 0.0025 \times 72 = 0.180 \text{ in.}^2 \text{ per linear foot}$$

The maximum spacing of reinforcement in walls is controlled by Section 7.6.5 of the ACI Code which provides that bars shall be spaced not farther apart than three times the wall thickness or more than 18 in. For this wall, the maximum spacing is $3 \times h = 3 \times 6 = 18$ in. Then

$$\text{vertical steel} = \text{No. 4 bars 18 in. on centers } (A_s = 0.13)$$

$$\text{horizontal steel} = \text{No. 4 bars 12 in. on centers } (A_s = 0.20)$$

both values being expressed as square inches per linear foot of wall.

Problem 13-3-A.* An 8-in.-thick reinforced concrete bearing wall is 15 ft high. It supports precast concrete girders 10 ft on centers and each has a service load reaction of 42 k. Of this amount 28 k is due to dead load and 14 k to live load. The girders have full bearing on the wall and the effective width of bearing (parallel to the wall face) may be taken as $7\frac{1}{2}$ in. Determine whether the wall is adequate for this loading and design the required reinforcement. Specification data: $f_c' = 3000$ psi and $f_y = 40,000$ psi.

13-4. Basement Walls

The identifying characteristic of a basement wall is that most of its height is below grade and it separates building space from earth in contact with its outside face. Such walls must be properly waterproofed and reinforced to provide for temperature variations as well as to resist bending stresses due to the thrust of the earth.

Basement walls may or may not be bearing walls, depending on the structural scheme employed in a particular building. With respect to the earth-retaining function, a basement wall may be considered as a slab spanning from column to column or as a slab with vertical tension reinforcement, the first and basement floor

slabs serving as the two reactions for the horizontal earth pressure. Because the basement height is generally less than the column spacing, the latter condition occurs most frequently.

Referring to Fig. 13-1a, the earth pressure is considered to be a horizontal triangular loading with a maximum value at the basement floor and decreasing in magnitude toward the top of the slab. The resultant of the earth pressure is represented by P and, when the surface of the retained earth at the top of the wall is horizontal, its magnitude may be determined by the formula

$$P = 0.286\,\frac{wh^2}{2}$$

in which w = the weight of the retained earth in pounds per cubic foot and h is the height of the retained earth in feet. (This use of

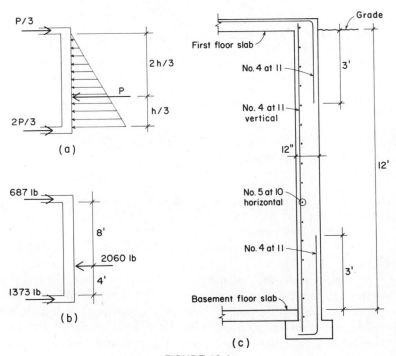

FIGURE 13-1.

the symbol h to denote height of retained earth or vertical distance between supports should not be confused with its meaning in Sec. 13-3, where it was employed to denote thickness of a bearing wall.) The direction of the resultant earth pressure is horizontal and acts at $\frac{1}{3}h$ from the bottom of the wall slab. The two reactions, or resisting horizontal forces, are $\frac{1}{3}P$ and $\frac{2}{3}P$, the forces resisted by the first floor and basement floor slabs, respectively. For this type of triangular loading the section of the wall slab at which the bending moment is maximum is $0.58 \times h$ from the top of the slab. A wall of this type is in reality a vertical slab with vertical reinforcement. If the wall and the floor slabs are placed at the same time, there is restraint at the two reactions. It is customary in computing the bending moment, however, to consider the slab as simply supported, thus erring on the side of safety. The maximum bending moment for this type of triangular loading is given by the formula

$$M = 0.128WL \quad \text{or} \quad M = 0.128Wh$$

where W is the total triangular load and L or h is the span length—the height or vertical distance between supports.

In many instances the required minimum thickness of 8 in. for basement walls, together with the minimum reinforcement ratio of $\rho = 0.0015$ for vertical steel, will satisfy the requirements for bending due to earth pressure. This is frequently the case in residential buildings.

One design procedure applicable to basement walls not supporting significant vertical superimposed loads is to assume a thickness, supply reinforcement in accordance with the minimum ratio, and then investigate this tentative design for adequacy in bending under the lateral load exerted by earth pressure. This procedure is followed in the example presented below. Local knowledge is required to establish a value for w, the weight of the retained earth. This can vary from an average value of approximately 100 lb/ft³ for dry sand to 120 lb/ft³ for wet sand or ordinary wet earth. The local building code should be consulted to ascertain whether mandatory values are specified.

Example. Design a reinforced concrete basement wall 12 ft 0 in. high between the basement and first-floor slabs that serve as sup-

ports to take the lateral earth pressure. Exterior finished grade is approximately level with the bottom of the first-floor slab, as indicated in Fig. 13-1c. No appreciable vertical load is carried by the wall because the span of the first-floor construction is parallel to the wall shown in the figure. Specification data: f'_c = 3000 psi, f_y = 40,000 psi, and the weight of the retained earth is to be taken as 100 lb/ft^3.

Solution: (1) The value of the total earth pressure for each linear foot of wall is

$$P = 0.286 \frac{wh^2}{2} = 0.286 \times \frac{100 \times 12 \times 12}{2} = 2060 \text{ lb}$$

In designing the wall, we consider a vertical strip of slab 12 in. wide; therefore the earth pressure on this strip is 2060 lb. This resultant force acts at $\frac{1}{3}h$, or 4 ft 0 in., from the basement floor slab.

The force resisted by the first floor slab is $\frac{1}{3} \times 2060 = 687$ lb, and the basement floor slab resists a force of $\frac{2}{3} \times 2060 = 1373$ lb. See Fig. 13-1b.

(2) The section of maximum bending moment in the wall slab is located $0.58 \times h$ from the top of the wall, or $0.58 \times 12 = 6.96$ ft, and the magnitude of the moment is

$$M = 0.128Wh = 0.128 \times 2060 \times 12 = 3170 \text{ ft-lb}$$

This represents the service live load moment, and because there is no service dead load moment

$$M_u = 1.7M_l = 1.7 \times 3170 = 5390 \text{ ft-lb}$$

and the required theoretical moment strength is

$$M_t = \frac{M_u}{\phi} = \frac{5390}{0.90} = 5988 \text{ ft-lb}$$

(3) Assume a thickness of 12 in. This gives a gross cross-sectional area A_g of 144 sq. in. for a 12-in. strip of wall 12 in. thick. If we take minimum ρ = 0.0015 as a trial reinforcement ratio,

$$A_s = \rho A_g = 0.0015 \times 144 = 0.216$$

which can be supplied by No. 4 bars spaced 11 in. on centers. This reinforcement is tentatively adopted.

(4) Following the procedure for investigation of rectangular sections (Section 4-14), the ultimate value of T (Fig. 4-4) is

$$T = A_s f_y = 0.216 \times 40,000 = 8640 \text{ lb}$$

Setting this value equal to the expression for C (Section 6-6) and solving for a, the depth of the rectangular stress block,

$$C = 0.85 f'_c ba = 8640 \text{ lb}$$

and

$$a = \frac{8640}{0.85 f'_c b} = \frac{8640}{0.85 \times 3000 \times 12} = 0.28 \text{ in.}$$

(5) Compute the theoretical moment strength of the 12-in. strip of wall by substituting in Formula (2) Section 6-6, taking d as 12 − $1\frac{1}{4}$ (concrete cover at inside face) = 10.75 in.

$$M_t = T \left(d - \frac{a}{2} \right) = 8640 \left(10.75 - \frac{0.28}{2} \right)$$

$$= 91,670 \text{ in-lb or } 7639 \text{ ft-lb.}$$

Because this value is greater than the required theoretical moment strength of 5988 ft-lb determined in Step 2, No. 4 bars on 11-in. centers are adopted for the vertical reinforcement. These bars are placed near the inner face of the wall.

(6) In addition to vertical bars, horizontal reinforcement must be used. The ACI Code requires that the horizontal reinforcement be a minimum of 0.0025 times the area of the reinforced section of the wall. Therefore $0.0025 \times 12 \times 12 = 0.36$ sq. in., and we use No. 5 bars spaced 10 in. on centers.

To provide for possible tensile stresses in the outer portion of the wall at the basement and first floor slabs, No. 4 bars are extended to the fourth points of height, as shown in Fig. 13-1c.

It will be found that neither shear nor development length of reinforcement are critical in this wall. This is commonly the situation for basement walls of this general type and proportions.

Problem 13-4-A. A basement wall 15 ft 0 in. high is supported by the basement and first-floor slabs against the lateral pressure of the exterior earth fill. Design the wall in accordance with the following specification data: $f'_c = 3000$ psi, $f_y = 40,000$ psi, and the weight of the earth fill is to be taken as 100 lb/ft^3.

13-5. Retaining Walls

Strictly speaking, any wall that sustains significant lateral soil pressure is a retaining wall. However, the term is usually used with reference to a so-called cantilever retaining wall, which is a freestanding wall without lateral support at its top. For such a wall the major design consideration is for the actual dimension of the ground-level difference that the wall serves to facilitate. The range of this dimension establishes the following different categories for the retaining structure:

> *Curbs.* Curbs are the shortest freestanding retaining structures. The two most common forms are as shown in Fig. 13-2, the selection being made on the basis of whether or not it is necessary to have a gutter on the low side of the curb. Use of these structures is typically limited to grade level changes of about 2 ft or less.
>
> *Short Retaining Walls.* Vertical walls up to about 10 ft in height are usually built as shown in Fig. 13-3. These consist of a concrete or masonry wall of uniform thickness. The wall thickness, footing width and thickness, vertical wall reinforcing, and transverse footing reinforcing are all designed for the

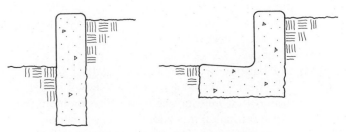

FIGURE 13-2. Typical concrete curbs.

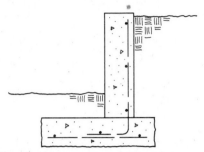

FIGURE 13-3.   Typical short retaining wall.

lateral shear and cantilever bending moment plus the vertical weights of the wall, footing, and earth fill.

Tall Retaining Walls.   As the wall height increases, it becomes less feasible to use the simple construction shown in Fig. 13-3. The overturning moment increases sharply with the increase in height of the wall. For very tall walls, one modification used is to taper the wall thickness. This permits the development of a reasonable cross section for the high bending stress at the base without an excessive amount of concrete. However, as the wall becomes really tall, it is often necessary to consider the use of various bracing techniques, as shown in the other illustrations in Fig. 13-4.

The design of tall retaining walls is beyond the scope of this book. They should be designed with a more rigorous analysis of the active soil pressure than that represented by the simplified equivalent fluid stress method. In addition, the magnitudes of forces in the reinforced concrete elements of such walls indicate the use of strength design methods rather than the less accurate working stress methods.

Under ordinary circumstances it is reasonable to design relatively short retaining walls by the equivalent fluid pressure method and to use the working stress method for the design of the elements of the wall. The following example illustrates this simplified method of design.

Example.   A short retaining wall is proposed with the profile

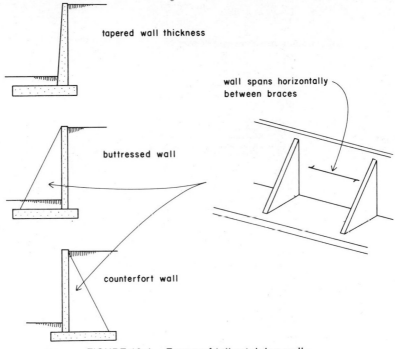

FIGURE 13-4.   Forms of tall retaining walls.

shown in Fig. 13-5. Investigate for the adequacy of the wall dimensions and select reinforcing for the wall and its footing. Use the following data.

Active soil pressure: 30 lb/ft^2 of height.
Soil weight: 100 lb/ft^3.
Maximum allowable soil pressure: 1500 psf.
Concrete strength: f'_c = 3000 psi.
Reinforcing: Grade 40 bars, f_y = 40,000 psi.

Solution: The loading condition used to analyze the stress conditions in the wall (above the footing) is shown in Fig. 13-6, and the analysis is as follows.

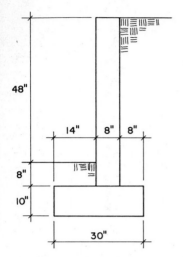

FIGURE 13-5.

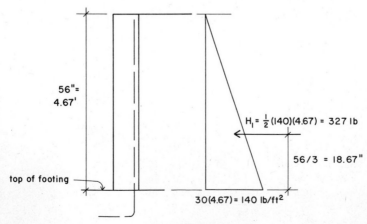

$H_1 = \frac{1}{2}(140)(4.67) = 327$ lb

$56/3 = 18.67''$

top of footing

$30(4.67) = 140$ lb/ft^2

FIGURE 13-6.

Maximum lateral pressure:

$$p = (30)(4.667 \text{ ft}) = 140 \text{ lb/ft}^2$$

Total horizontal force:

$$H_1 = \frac{(140)(4.667)}{2} = 327 \text{ lb}$$

Moment at base of wall:

$$M = (327) \left(\frac{56}{3}\right) = 6104 \text{ lb-in.}$$

For the wall, we assume an approximate effective d of 5.5 in. The tension reinforcing required for the wall is thus

$$A_s = \frac{M}{f_s jd} = \frac{6104}{(20,000)(0.9)(5.5)} = 0.061 \text{ in.}^2/\text{ft}$$

This may be provided by using No. 3 bars at 20-in. centers, which gives an actual A_s of 0.066 in.^{2}/ft. Since the embedment length of these bars in the footing is quite short, they should be selected conservatively and should have hooks at their ends for additional anchorage.

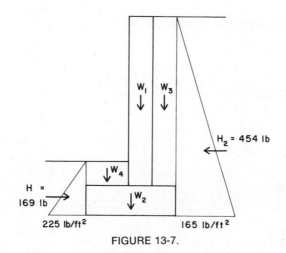

FIGURE 13-7.

TABLE 13-1.   Determination of the Eccentricity of the Resultant Force

Force (lb)		Moment arm (in.)	Moment (lb-in.)
H_2	454	22	+9988
w_1	466	3	−1398
w_2	312	0	
w_3	311	11	−3421
w_4	78	8	+624
$\Sigma_w = 1167$ lb		Net moment =	+5793 lb-in.

The loading condition used to investigate the soil stresses and the stress conditions in the footing is shown in Fig. 13-7. In addition to the limit of the maximum allowable soil bearing pressure, it is usually required that the resultant vertical force be kept within the kern limit of the footing. The location of the resultant force is therefore usually determined by a moment summation about the centroid of the footing plan area, and the location is found as an eccentricity from this centroid.

Table 13-1 contains the data and calculations for determination of the location of the resultant force that acts at the bottom of the footing. The position of this resultant is found by dividing the net moment by the sum of the vertical forces, as follows:

$$e = \frac{5793}{1167} = 4.96 \text{ in.}$$

For the rectangular footing plan area, the kern limit will be one-sixth of the footing width or 5 in. The resultant is thus within the kern, and the combined soil stress may be determined by the stress formula as follows:

$$p = \frac{N}{A} \pm \frac{M}{S}$$

in which N is the total vertical force

A is the plan area of the footing

M is the net moment about the footing centroid

S is the section modulus of the rectangular footing plan area, which is determined as follows:

$$S = \frac{bh^2}{6} = \frac{(1)(2.5)^2}{6} = 1.042 \text{ ft}^3$$

The limiting maximum and minimum soil pressures are thus determined as follows:

$$p = \frac{N}{A} \pm \frac{M}{S} = \frac{1167}{2.5} \pm \frac{5793/12}{1.042} = 467 \pm 463$$

$$= 930 \text{ lb/ft}^2 \text{ maximum and } 4 \text{ lb/ft}^2 \text{ minimum}$$

Since the maximum stress is less than the established limit of 1500 lb/ft^2, vertical soil pressure is not critical for the wall. For the horizontal force analysis, the procedure varies with different building codes. The criteria given in this example for soil friction and passive resistance are those in the *Uniform Building Code* (Ref. 8) for ordinary sandy soils. This code permits the addition of these two resistances without modification. Using this data and technique, the analysis is as follows:

Total active force: 454 lb, as shown in Fig. 13-7.

Friction resistance [(friction factor)(total vertical dead load)]:

$$(0.25)(1167) = 292 \text{ lb}$$

Passive resistance: 169 lb, as shown in Fig. 13-7.

Total potential resistance:

$$292 + 169 = 461 \text{ lb}$$

Since the total potential resistance is greater than the active force, the wall is not critical in horizontal sliding.

As with most wall footings, it is usually desirable to select the footing thickness to minimize the need for tension reinforcing due to bending. Thus shear and bending stresses are seldom critical, and the only footing stress concern is for the tension reinforcing. The critical section for bending is at the face of the wall, and the loading condition is as shown in Fig. 13-8. The trapezoidal stress

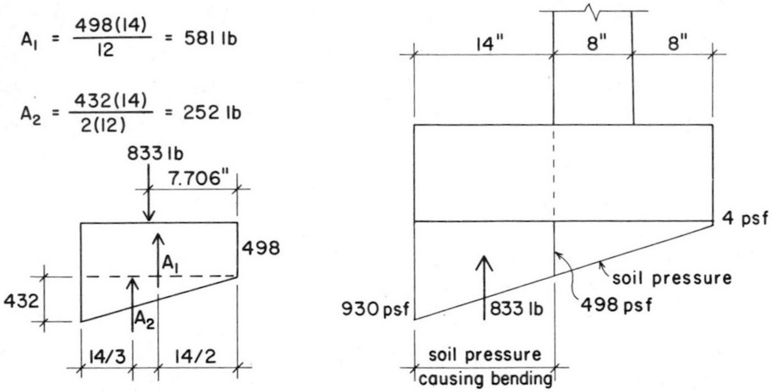

$$A_1 = \frac{498(14)}{12} = 581 \text{ lb}$$

$$A_2 = \frac{432(14)}{2(12)} = 252 \text{ lb}$$

FIGURE 13-8.

distribution produces the resultant force of 833 lb, which acts at the centroid of the trapezoid, as shown in the illustration. Assuming an approximate depth of 6.5 in. for the section, the analysis is as follows:

Moment:

$$M = (833)(7.706) = 6419 \text{ lb-in.}$$

Required area:

$$A_s = \frac{M}{f_s\, jd} = \frac{6149}{(20,000)(0.9)(6.5)} = 0.055 \text{ in.}^2/\text{ft}$$

This requirement may be satisfied by using No. 3 bars at 24-in. centers. For ease of construction, it is usually desirable to have the same spacing for the vertical bars in the wall and the transverse bars in the footing. Thus, in this example, the No. 3 bars at 20-in. centers previously selected for the wall would probably also be used for the footing bars. The vertical bars can then be held in position by wiring the hooked ends to the transverse footing bars.

Although bond stress is also a potential concern for the footing bars, it is not likely to be critical as long as the bar size is relatively small (less than a No. 6 bar or so).

Reinforcing in the long direction of the footing should be determined in the same manner as for ordinary wall footings. As discussed in Section 13-2, we recommend a minimum of 0.15% of the cross section. For the 10-in.-thick and 30-in.-wide footing, this requires

$$A_s = (0.0015)(300) = 0.45 \text{ in.}^2$$

We would therefore use three No. 4 bars with a total area of $(3)(0.2) = 0.6$ in.^{2}.

In most cases designers consider the stability of a short cantilever wall to be adequate if the potential horizontal resistance exceeds the active soil pressure and the resultant of the vertical forces is within the kern of the footing. However, the stability of the wall is also potentially questionable with regard to the usual overturn effect. If this investigation is considered to be necessary, the procedure is as follows.

The loading condition is the same as that used for the soil stress analysis and shown in Fig. 13-7. As with the vertical soil stress analysis, the force due to passive soil resistance is not used in the moment calculation, since it is only a potential force. For the overturn investigation, the moments are taken with respect to the toe of the footing. The calculation of the overturning and dead load restoring moments are shown in Table 13-2. The safety factor against overturn is determined as

$$SF = \frac{\text{restoring moment}}{\text{overturning moment}} = \frac{20,686}{9988} = 2.07$$

The overturning effect is usually not considered to be critical as long as the safety factor is at least 1.5.

Table 13-3 gives design data for short reinforced concrete retaining walls varying in height from 2 to 6 ft. Table data has been developed using the procedures illustrated in the example. Details and criteria for the walls are shown in Fig. 13-9. Note that the illustration shows two necessary conditions. The first concerns the profile of the ground surface behind the wall. If this has a significant slope, there will be an increase in the active soil pressure similar to that due to a surcharge. Table designs are

TABLE 13-2. Analysis for Overturning Effect

Force (lb)	Moment arm (in.)	Moment (lb-in.)
Overturn:		
H_2 454	22	9988
Restoring moment:		
w_1 466	18	8388
w_2 312	15	4680
w_3 311	26	8086
w_4 78	7	546
		Total: 20,686 lb-in.

based on consideration of an essentially flat profile, although a very minor slope (up to 5 : 1, as shown) will not cause significant increase in pressure. The second requirement is that care be taken to avoid the possibility of highly saturated soil behind the wall. This should be avoided by using a reasonably permeable fill and by placing drains in the wall as shown.

Problem 13-5-A. Design a short retaining wall similar in form to that shown in Fig. 13-9. Referring to the figure, $H = 4.5$ ft. Use $f'_c = 2$ ksi [13.8 MPa] and Grade 40 bars with $f_y = 40$ ksi [276 MPa] and $f_s = 20$ ksi [138 MPa].

13-6. Shear Walls

Reinforced concrete shear walls represent some of the strongest and stiffest elements for bracing of a building against lateral forces. (See Fig. 13-10.) In many cases they also function as bearing walls. The complete design of a shear wall involves considerations of sliding, overturn effects, transfer of loads from supported elements, and analysis for load distribution as well as the investigation of the shear and flexural stresses in the wall. Various combinations of dead, live, and lateral loads must often be considered. A complete development of this topic is well beyond the scope of this book.

The ACI Code provides special material in its Appendix A

TABLE 13-3. Short Concrete Retaining Walls[a]

Wall height H (ft)	Wall and footing dimensions (ft-in.)				Reinforcing				Actual maximum soil pressure (lb/ft²)
	w	h	t	A	1	2	3	4	
2	1–6	0–6	0–6	0–4	No. 3 at 30	—	—	2 No. 3	750
3	2–0	0–8	0–6	0–6	No. 3 at 24	1 No. 4	—	2 No. 4	800
4	2–6	0–10	0–8	0–8	No. 3 at 20	2 No. 4	No. 3 at 20	3 No. 4	950
5	3–4	1–0	0–9	1–1	No. 4 at 24	3 No. 4	No. 4 at 24	4 No. 4	900
6	4–4	1–3	0–10	1–4	No. 4 at 18	4 No. 4	No. 4 at 18	4 No. 5	925

[a] See Fig. 13-9 for reference. Design f'_c = 2 ksi [13.8 MPa], Grade 40 bars with f_y = 40 ksi [276 MPa].

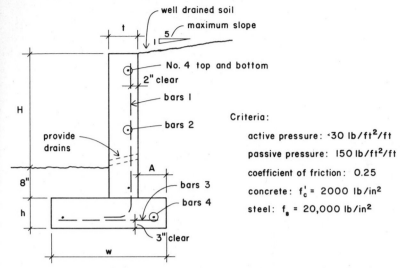

FIGURE 13-9. Reference figure for Table 13-3.

regarding seismic design in general, including design of shear walls. In addition, codes in regions where seismic design is a critical concern generally provide considerable criteria for design of shear walls. For a general discussion of shear walls and some examples of their design, the reader is referred to *Simplified Building Design for Wind and Earthquake Forces* (Ref. 12).

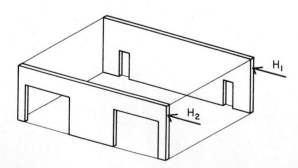

FIGURE 13-10. Use of walls to resist lateral loads on a building.

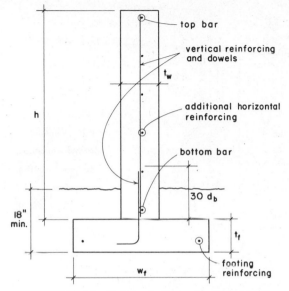

FIGURE 13-11.   Reference figure for Table 13-4.

13-7.   Freestanding Walls

Where walls are not required to connect to the ceiling or roof at their tops, freestanding walls are sometimes used inside buildings as partitions. In these cases the only real structural design criteria may be the consideration of lateral seismic force, where such is

TABLE 13-4.   Freestanding Walls (see Fig. 13-11)

h (ft)	t_w (in.)	w_f (in.)	t_f (in.)	Top and bottom bars	Additional horizontal reinforcing	Vertical reinforcing and dowels	Footing reinforcing
4	6	18	8	No. 5	2 No. 4	No. 4 at 18	2 No. 3
5	6	18	8	No. 5	3 No. 4	No. 4 at 18	2 No. 3
6	6	22	8	No. 6	4 No. 5	No. 4 at 18	2 No. 3
7	8	26	8	No. 6	5 No. 5	No. 4 at 16	2 No. 4
8	8	30	8	No. 6	6 No. 5	No. 4 at 12	3 No. 4
9	8	34	10	No. 7	8 No. 5	No. 5 at 12	3 No. 4
10	9	38	10	No. 7	9 No. 5	No. 5 at 10	3 No. 5

appropriate. When occurring outdoors, there will always be a concern for wind forces, which may be quite serious for tall walls where windstorm conditions are severe. In most ordinary situations, these walls are designed quite empirically, based on someone's judgment or experience. Walls of moderate height usually take the form shown in Fig. 13-11, with a single layer of reinforcing in the wall and a footing reinforced only in the long direction. Table 13-4 gives recommended data for freestanding walls of the form shown in Fig. 13-11. These are in general reasonably conservative, except where severe wind, high seismic force, or very poor soil conditions occur. In such cases, a full engineering investigation should be done to verify the adequacy of the data.

References

IIIIIIIIIIIIIIIIIIIIIIIIIIIIIIIIIIIIIIIIIIIII

1. *Building Code Requirements for Reinforced Concrete,* ACI 318-77, American Concrete Institute, Detroit, (Commonly called the ACI Code.)

2. *Building Code Requirements for Reinforced Concrete,* ACI 318-63, American Concrete Institute, Detroit, 1963. (The last edition with a complete development of the working stress method.)

3. *CRSI Handbook,* Concrete Reinforcing Steel Institute, Schaumburg, IL, 1982.

4. *Reinforced Concrete Fundamentals,* 4th ed., Phil M. Ferguson, Wiley, New York, 1979.

5. *Simplified Engineering for Architects and Builders,* 6th ed., Harry Parker and James Ambrose, Wiley, New York, 1983.

6. *Simplified Design of Building Foundations,* James Ambrose, Wiley, New York, 1981.

7. *Simplified Building Design for Wind and Earthquake Forces,* James Ambrose and Dimitry Vergun, Wiley, New York, 1980.

8. *Uniform Building Code,* 1982 ed., International Conference of Building Officials, Whittier, CA.

9. *Notes on ACI 318-77,* Portland Cement Association, Skokie, IL, 1980.

10. *Reinforced Concrete Design Handbook,* Working Stress Method (ACI Publication SP-3), American Concrete Institute, Detroit, 1965.

11. *ACI Detailing Manual* (ACI Publication SP-66), American Concrete Institute, Detroit, 1980.

12. *Simplified Design of Building Foundations,* James Ambrose, Wiley, New York, 1981.

Index

IIIIIIIIIIIIIIIIIIIIIIIIIIIIIIIIIIIIIIIIIIIII

Abbreviations, 5, 6, 7
ACI Code, 3, 56
Admixtures, 13
  air-entraining, 13
Aggregates, 11
  coarse, 12
  fine, 11
  grading of, 11
  lightweight, 12
  proportioning of, 20
  size, 11
Air-entraining cement, 10, 13
Allowable deflection, 114
Allowable stress:
  bearing, 212
  bending, 55
  compression, 187
  shear, 74
Alternate design method, ACI Code, 3,
  56
American Concrete Institute, 3, 56
American Society for Testing and
  Materials, 10
Anchorage of reinforcement, 94, 208
Areas of reinforcing bars, inside cover,
  in slabs, 134

Balanced reinforcement, 58
  strength design, 66
  working stress, 58
Bars, areas of, inside cover development:
  in columns, 107, 208
  in continuous beams, 105

in footings, 208
in pedestals, 208
in simple beams, 101
spacing in beams, 82
Basement walls, 227
Beams:
  bending in, 34, 52
  cantilever, 36
  compression reinforcement in, 136
  continuous, 38, 159
  deflection, 114
  depth, 113
  design, 55, 65, 110, 118
  doubly reinforced, 136
  effective depth, 53
  internal resisting moment, 52
  investigation of, 52
  rectangular, 52, 65
  shape, 111
  shear in, 73, 76
  spacing of bars in, 47
  strength design of, 65
  T-beams, 126
  types of, 36
  under-reinforced, 64
  width of, 112
  working stress design of, 60
Bearing walls, 222, 224
Bending, theory of, 52
Bending formulas:
  strength design, 65
  working stress, 56
Bending of reinforcement, 47

Bending stresses, 55
Bond stress, 94

Cantilever beam, 37
Cement:
　air-entraining, 10, 13
　high-early-strength, 10
　portland, 1, 10
Coarse aggregate, 11
Code, ACI, 3, 56
Column footing, 197
Column pedestal, 208
Columns, 182
　axial load, 182
　eccentric load, 182
　effective length, 194
　footings for, 197
　interaction of axial load and moment,
　　182
　length, 194
　minimum eccentricity, 182
　pedestals for, 208
　round, 184, 194
　spiral, 184
　tied, 184
　types of, 184
Compression reinforcement in beams,
　136
Compression test, 31
Concentrated loads on beams, 36
Concrete:
　air-entrained, 10, 13
　creep of, 17
　curing of, 23, 27
　joists, 171
　lightweight, 12
　mixing of, 18
　modulus of elasticity of, 16
　pedestal, 208
　properties of, 9
　reinforced, 1
　slump test, 29
　strength of, 15
　testing of, 29
　waffle construction, 174
Consistency, 118

Continuous beams, 40
Conversion of units, inside cover
Cover of reinforcement, 46
Creep, 17
Curing, 23, 27

Dead load, 148
Deflection, 114
Design aids, 40, 181
Design controls, 24
Design methods:
　strength, 3, 44
　working stress, 3, 43
Design procedure:
　one-way slab, 133, 155
　rectangular beam, 118
　T-beam, 127
Design values for beams, 55
Development length, 94
　in columns, 107, 208
　in continuous beams, 105
　in footings, 208
　in pedestals, 208
　in simple beams, 101
Diagonal tension, 74
Doubly reinforced beam, 134
Durability, 17

Effective depth, of beams, 53
Effective width, of T-beam flange, 112,
　127
Exposure, degrees of, 17, 28
Extreme fiber stress, 54

Factored load, 44
Fine aggregate, 11
Fixed beam, 36
Flange width, of T-beam, 112, 127
Flat plate construction, 178
Flat slab construction, 178
Flexural formulas:
　strength design, 65
　working stress, 56
Floor systems, 147
Footing:
　column, 197

wall, 214
Formwork, 25
Foundation beds, allowable bearing
    capacity, 195
Foundations, 195

Girder, design of, 167
Grade of steel, 14

High-early-strength cement, 10
Hooks, 47
    equivalent development length of, 97
    standard, 47, 99
Hydration, 9

Independent column footing, 197
Ingredients, proportioning of, 20
Installation of reinforcing, 32
Internal resisting moment, 52
Investigation of beams, 34

Joists, concrete, 171

k factors for beams, 60

Laitance, 27
Lapped splice, 108
Lightweight aggregates, 12
Lightweight concrete, 12
Live loads, 150
    reduction of, 150
Load factors, 44
Loads:
    dead, 148
    factored, 44
    live, 150
    service, 44
Long-time deflection, 114

Minimum dimensions for concrete
    members, 48
Minimum reinforcement, 50
Mixing concrete, 18
Modulus of elasticity, 16

Neutral axis, 54

Neutral surface, 54
Nomenclature, 7

One-way slab, 131, 152

Pedestal, 208
Peripheral shear, 73, 199
Placement of concrete, 26
Portland cement, 1, 10
Properties of concrete, 9
Properties of reinforcing bars, inside
    cover
Proportioning concrete mixes, 20
Punching shear, 73, 199

Reactions, 36
Rectangular beam, 56, 65
Rectangular stress block, 54
Reduction of live load, 150
Reinforcement:
    anchorage of, 94
    areas of, inside cover
    balanced, 58, 66
    compressive, 136
    concrete cover for, 46
    grades of, 14
    shrinkage, 50, 132
    temperature, 50, 132
    welded wire, 14
Retaining wall, 222, 232
Round column, 184, 194

Sand, 11
Segregation, 18
Service load, 44
Shear, 73
Shear reinforcement, 75, 78
    design of, 83, 90
Shop drawings, 32
Slabs:
    one-way, 131
    ribbed, 171, 174
    thickness of, 114
    two-way, 177
    waffle, 174
Slenderness, of columns, 194

Slump test, 29
Spacing:
  of bars, 47
  of stirrups, 82
Specified compressive stress, 54
Spiral column, 184
Splices, 108
Steel reinforcement, grades of, 14
Stirrups, 78
  spacing of, 82
Strength, of concrete, 15
Strength design method, 3, 44, 55, 65
Strength reduction factor, 44
Symbols, 8

T-beam, 126
Temperature reinforcement, 132
Testing concrete, 29
Tied column, 184
Two-way slab, 177

Ultimate strength design, 44
Under-reinforced beam, 64

Units of measurement, inside cover, 4

Vertical shear, 73, 76

Waffle construction, 174
Wall footing, 214
Walls, 222
  basement, 222, 227
  bearing, 222, 224
  free-standing, 244
  retaining, 222, 232
  shear, 222, 241
Water, 11
Water-cement ratio, 19
Water tightness, 19
Web reinforcing, 73
Weights of materials, 148
Welded wire fabric, 14
Workability, 18
Working stress method, 3, 43, 55, 60

Yield strength, of reinforcing, 14

Properties of Standard Reinforcing Bars

Size	Nominal diameter (in.)	Nominal diameter (mm)	Nominal area (in.²)	Nominal area (mm²)	Nominal perimeter (in.)	Nominal perimeter (mm)	Weight (lb/ft)	Weight (kg/m)
3	0.375	9.52	0.11	71	1.178	29.92	0.376	0.560
4	0.500	12.70	0.20	129	1.571	39.90	0.668	0.994
5	0.625	15.88	0.31	200	1.963	49.86	1.043	1.552
6	0.750	19.05	0.44	284	2.356	59.84	1.502	2.235
7	0.875	22.22	0.60	387	2.749	69.82	2.044	3.042
8	1.000	25.40	0.79	510	3.142	79.81	2.670	3.973
9	1.128	28.65	1.00	645	3.544	90.02	3.400	5.060
10	1.270	32.26	1.27	819	3.990	101.35	4.303	6.404
11	1.410	35.81	1.56	1006	4.430	112.52	5.313	7.907
14	1.693	43.00	2.25	1452	5.320	135.13	7.650	11.380
18	2.257	57.33	4.00	2581	7.090	180.09	13.600	20.240

ID9919871